The Basics of Agile Product Development

Joachim Pfeffer

is a management consultant in lean and agile product development. After more than twenty years in product development (software, electronics, and mechanics) and ten years of consulting experience in development and service processes, Joachim Pfeffer supports companies with the introduction of lean and agile concepts in embedded and mechanical engineering. Some of Joachim's recent projects include the development of camshafts, fuel cells, train compressors, control units, and semiconductors with Scrum.

Joachim Pfeffer

The Basics of Agile Product Development

An Introduction to Scrum, Kanban, and Lean Development

Bibliographic information published by the Deutsche Nationalbibliothek
The Deutsche Nationalbibliothek lists this publication in the Deutsche Nationalbibliografie; detailed bibliographic data are available in the Internet at http://dnb.d-nb.de

The designations used in this book as well as brand names and product designations of the respective companies are generally subject to trademark, brand or patent protection. All information and programs in this book have been checked with the utmost care. However, neither the authors nor the publisher can be held liable for any damages arising in connection with the use of this book.

First edition (internal: v3.0)
Copyright © 2024   peppair GmbH
Oberweiler 2, 88239 Wangen im Allgäu, training@peppair.com

Editor:	Jessica Thamm, www.natives.de
Illustrations:	Joachim Pfeffer
Typeset:	Joachim Pfeffer

Printed by Libri Plureos GmbH, Friedensallee 273, 22763 Hamburg

Printed in Germany
ISBN: 978-3-947487-22-6

Content

People and Teams   169

Literature   189

Foreword

This is the translation of the german second edition published in 2022. In 2015, I began writing down my experiences with the introduction of Kanban in product development, i.e., beyond just software development. Over the years, the planned scope expanded from just a Kanban book to a basic work on the agile development of mechatronic systems because Scrum has increasingly become the focus of my projects. However, this increased the risk that writing the planned book would take forever. I have been conducting training courses on agile approaches on a larger scale since 2016, and this book includes the basic topics. I use it as a handout for my training courses, but it is also intended to serve as a stand-alone basic book that brings together the ideas of Kanban, Scrum, and lean development. The extended book on agile product development has now been published by Carl Hanser Verlag and is called *Produktentwicklung – Lean & Agile*.

My aim with this book is to provide the basic knowledge on these very interesting approaches and make a small contribution to better products and market opportunities, as well as spreading more joy in the product development world. I am therefore making the contents of this book available for non-commercial use under a Creative Commons license.

Joachim Pfeffer, September 2024

Introduction

Agile development and the Scrum framework are attracting a lot of attention both within and outside the field of pure software development. For some, it is the consistent implementation of the only reliable project management method, for others it is just a hype that brings no new insights or benefits.

The fact is that Scrum has become the de facto standard in software development and has achieved considerable success in terms of development speed, predictability, and quality. In my opinion, the occasional opinion that this software development success story cannot be transferred to other aspects fall short of the mark.

The success of agile development is only partly based on practicing a specific methodology. Many advantages are only achieved through a gradual change in mindset and culture – factors that are independent of industry, product, and technology.

Becoming agile therefore means far more than just successfully implementing a method. It is a paradigm shift, a departure from much of the tradition and culture that has matured around product development and project management over the past decades and centuries. A change that is significantly more difficult, more complex and more painful than the application of a new method. Contrary to the expectations of many managers, the changes do not affect the developers as much as the managers themselves, and even more so the higher up they are in the organizational hierarchy.

This change will always begin with the introduction of Scrum, Kanban or other agile approaches. The specific approach evolves over time, as do the associated changes. This book explains Scrum and Kanban, and it describes the connections between Lean Production and Lean Development. It can be a starting point on your journey to agility and change.

If you don't like change,
you're going to like irrelevance even less.
Eric Shinseki

Rethinking

The world is changing

Why agile product development? Hasn't everything been working out without Scrum? Is agile development the solution to all problems or just another new hype? The answers to these questions depend on how complicated or complex the environment – in which your organization operates – is. I will therefore introduce you to two thinking models that you can use to form your own opinion. Based on this observation, you then decide how great the need is to immerse yourself in a fast and flexible world of product development with agile thinking. Of course, even with these models there are no definitive answers, no black and no white. As the saying goes, "All models are wrong. Some are useful."

Cynefin

The Cynefin model by systems and organization researcher Dave Snowden divides the environment or system in which you find yourself in into four basic types (Figure 1), and discusses the appropriate solution strategy for each type (Snowden & Boone, 2007).

- The first environment is the **obvious** environment. This is the world of so-called best practices, with proven, standardized solution patterns. The only challenge here is to classify the problem correctly; it must be categorized precisely to be able to choose the correct pattern of action. According to Snowden, the solution strategy consists of *sense, categorize, respond*. Examples of solution patterns are a flyback diode on a relay, a nylon-insert lock nut on a bolt connection, or folding down the sun visor in a car when the sun is low in the sky.
- The world of the obvious is limited to small, local solutions and makes no relevant contribution to our discussion in prod-

uct development given that products are systems that consist of many parts and many obvious small environments. However, the interaction of these individual parts is no longer obvious, but according to Snowden **complicated**. The essential feature of a complicated system is a clear cause-and-effect relationship. It is deterministic, so experts can understand it and predict its behavior. In complicated systems, there is no longer a best practice that provides an off-the-shelf solution. Instead, there are several good solutions that have been developed by analyzing the system and based on existing experience. It is the world of good practices. An important aspect of a complicated system is that once it can be fully understood, it is also possible to create a plan that works reliably (*sense, analyse, respond*). All that remains is to monitor the execution of the plan

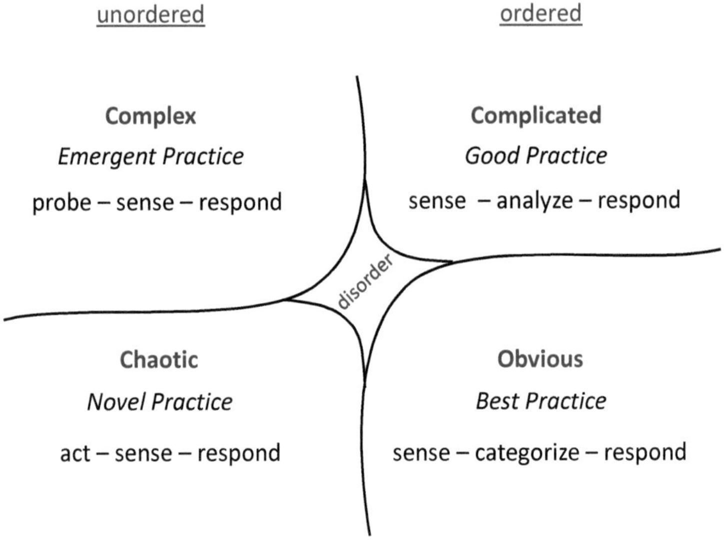

Figure 1: Cynefin model

and ensure that it is adhered to. This is the world of classic project management. Snowden describes the two right-hand quadrants, obvious and complicated, as ordered. It is the ordered world in which solution patterns and plans work.

- We leave this world with the top left context that stands for **complex** systems. Just like complicated systems, these have clear cause-and-effect relationships, however those are no longer recognizable to the problem solver. Although the system is deterministic, it does not behave that way when viewed from the outside because it cannot be understood. Here it becomes clear that the difference between complicated and complex cannot be defined by the system, but lies in the expertise of the observer: While the clockwork mechanism of my watch is a complex system for me, my watchmaker will see it as a complicated system. What practices and solution strategies are there in the complex environment in which empirical practices cannot be used? We are in a world of emergent practices, where the approach and solution emerge, grow, and change with new insights. In complex environments, the one who constantly interacts with the system to be able to adapt the plan accordingly will reach the goal. The action strategy is *probe, sense, respond*. These three steps are repeated several times on the way to a solution. The smaller the loops, the closer you stay to the – initially unknown – solution. This is the world of empirical plans and processes, the world of agile approaches.
- I will only briefly touch upon the last quadrant: **chaotic**. Even if many development projects feel like this, they are usually not chaotic, but in fact complex. In chaotic environments, there are no longer any cause-effect relationships, and *probe, sense, respond* does not lead to the goal, as every probe produces a different result. In this environment, you can only survive

with creativity and new practices. It is the world of novel practice, and the action strategy is *act, sense, respond.* Action is taken directly, no longer just checked and the plan is then adapted accordingly.

It is important to understand the Cynefin framework as a mental model. Do not use it as an assessment for your own situation as the solid lines drawn in the diagram are grey areas in practice. Often you can only guess the context you are currently in. Snowden describes this uncertainty in the context as *disorder.* Additionally, environments and systems often consist of partial aspects that can be assigned to different contexts in the Cynefin framework.

From the perspective of the Cynefin model, all industries are moving at different speeds from complicated to complex because of digitalization, globalization, the networking of markets, exponentially growing knowledge and more intense communication, and the associated increasing dynamics. Where is your organization and your product? Does the use of agile methods make sense for you? It might be worth considering a similar model by Ralph Stacey for your product development considerations.

Stacey Matrix

Ralph Stacey, a British organizational theorist, arranged leadership and decision-making concepts in a two-dimensional representation. On one axis, he demonstrates the degree of certainty and on the other the degree of agreement (Stacey, 2002). This basic layout is often used to explain the complicated and complex categories in product development. The meaning of the categories and the solution strategies can be taken from Dave Snowden.

The Stacey Matrix modified for product development shows the familiarity with the intended technology on the horizontal axis, which corresponds to the degree of certainty. At the origin, the technology is familiar and known, and towards the right, the uncertainty

                                        Rethinking

increases. In the same way, the familiarity or stability of the requirements, i.e., the degree of agreement, is plotted on the vertical axis.

The **simple** (corresponds to obvious in Cynefin), **complicated**, **complex**, and **chaotic** are depicted in the Stacey Matrix according to Figure 2.

Complexity in product development therefore arises from the technology, the requirements, or a combination of both. The various providers of social media, for example, work with known technologies, but must find out and verify the requirements through constant interaction with the market. In other words, if you already knew today what you would have to develop in order to be bought by Meta for a billion euros next year, you would not be reading this book now.

The moon landing can serve as a counter-example. In 1961, John F. Kennedy formulated the requirement to take man to the moon

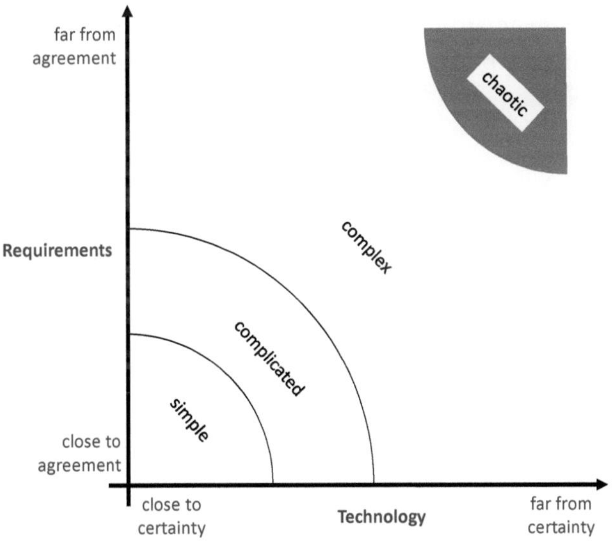

Figure 2: Stacey matrix

and bring him back again in one piece. This primary requirement was relatively clear. However, the required technology first had to be developed. A complex project.

Feedback

Learning, improving, adapting, and reacting – all these requires closed control loops. Feedback loops are necessary at various levels, especially in complex environments. In product development, there are control loops for clarifying requirements, optimizing working methods, and controlling the project progress on the time axis. However, the models by Dave Snowden and Ralph Stacey do not make a clear statement about the types of feedback loops, so I will briefly look at the control loop for validating requirements.

If you would like to clarify requirements, you get the best feedback by simply putting the product to use. An example is given in Figure 3: Someone intends to solve a business issue with a product. The aim is to generate revenue or avoid costs. A specification is created for this purpose, which is then implemented in a product.

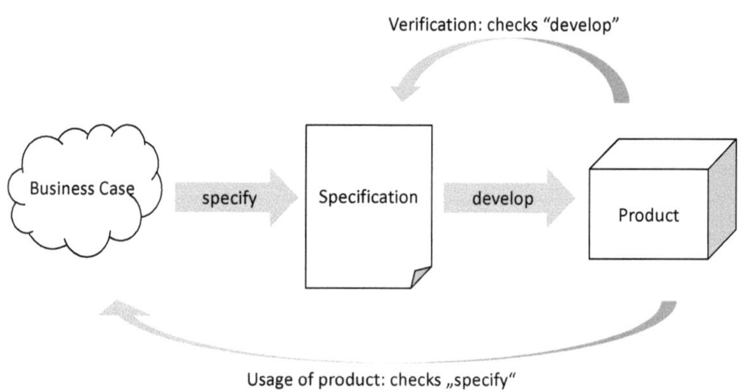

Figure 3: Feedback

The second transition, from specification to product, is development. It is ensured by verifying the product with the specification. However, it is much more difficult to ensure the transition from business to specification. Even if this transition works perfectly, the business can change during development – and therefore the specification would no longer match the business after development has been complete. The best way to secure this first transition is to transfer the product to the business (see lower arrow). In typical development projects, this theoretically only happens once at the end of the project, when time and money have been used up. Adjustments after feedback from the business are no longer possible. In recent decades, this setup has often meant there are two losers: the customer *and* the developer.

Of course, the issue is more complex in practice. In the automotive industry, for example, this feedback loop is executed several times in sample phases. For a *probe, sense, respond* approach in a complex world, however, these must be significantly faster. If you intend to illustrate the implementation of agile approaches in the development of physical products, this simple diagram is an important catalyst for discussion and definition. What results can you generate in a period of a few weeks, *and* who can provide sound feedback on these results?

Agile development

Overview

After many software projects failed in the 1980s and 1990s due to their complexity, a movement known as agile development has been about since the 1990s. It claims that the waterfall-like phase-driven process models in place for a long time are not suitable for development projects if the product requirements or the technology used, or both, are unknown. Agility meets these challenges with an emergent approach, an iterative-incremental development.

While prototypes and feedback loops have been used rather intuitively for unknown technologies in the past, it was long believed that the requirements for the product to be developed could be clearly defined at the start of the project. There are various definitions of agile product development. I would describe it using the following four attributes:

- Self-organized, team-based work
- Iterative-incremental development
- Clocked work, frequent delivery
- Willingness to continuously improve the organization

Various agile approaches have emerged since the 1990s. The best known is Scrum, which is now considered the actual standard in software development.

The Agile Manifesto

The Agile Manifesto, originally known as the Manifesto for Agile Software Development, was created at a conference in Utah in 2001 by 17 leading experts for new approaches in software development who agreed on what agile software development meant to them. This document is still the reference for assessing one's own way of working today and has lost none of its relevance and importance even after almost 25 years. The Agile Manifesto is divided into four values or pairs of values and 12 principles, which I would like to briefly introduce (Beck et al., 2001).

The four pairs of values in the agile manifesto are

We are uncovering better ways of developing software by doing it and helping others do it. Through this work we have come to value:
- *Individuals and interactions over Processes and tools*
- *Working software over Comprehensive documentation*
- *Customer collaboration over Contract negotiation*
- *Responding to change over Following a plan*

That is, while there is value in the items on the right, we value the items on the left more.

If you are reading this for the first time, you may be confused, because the values to the right of the word over are probably of central importance in your previous work. Even if many people interpret it differently: The Agile Manifesto does not deny the importance of these values, but affirms their significance in the last sentence.

It is interesting to note that the values alone cannot have any meaning, but can only be meaningfully applied as pairs of values. For example, consider the value honesty in isolation. Honesty sounds good, but cannot be a value on its own. After all, if you honestly tell someone you don't agree with what they think, this will probably have a negative impact on the social fabric between you both. The counter-value in this case would be respect. As a rule, you will subconsciously try to balance honesty and respect when dealing with disagreeable people. However, one aspect will always dominate slightly more than the other.

If we look at the Agile Manifesto from this values perspective, we no longer read – as is unfortunately often the case – that comprehensive documentation is no longer important, but that comprehensive documentation, as an isolated value, leads to a dead end without functioning software. The parallel growth of both values is therefore essential in order to be balanced out. Consequently, neither side should be promoted alone.

I often see how the Agile Manifesto is used to argue against agility because its very compact wording easily allows for misinterpretation. Professionally more careful colleagues who deal with quality assurance, functional safety, and process maturity models sometimes use the manifesto as proof that agile development cannot work in their industry. Discussions based on the manifesto can therefore lead to irritation under certain circumstances. The pair of values approach helps you to interpret these short statements in a better way. How do you interpret these values in your industry, your organization, and your business?

In addition to the four sets of values, the manifesto lists 12 principles that serve as guidelines for the organization of your day-to-day work. They are not based on abstract approaches, but are a collection of good practices from experienced software developers and project managers. The principles help everyone involved to reflect on or change their current approach.

The Agile Manifesto is based on 12 principles:

1. *Our highest priority is to satisfy the customer through early and continuous delivery of valuable software.*
2. *Welcome changing requirements, even late in development. Agile processes harness change for the customer's competitive advantage.*
3. *Deliver working software frequently, from a couple of weeks to a couple of months, with a preference to the shorter timescale.*
4. *Business people and developers must work together daily throughout the project.*
5. *Build projects around motivated individuals. Give them the environment and support they need, and trust them to get the job done.*
6. *The most efficient and effective method of conveying information to and within a development team is face-to-face conversation.*
7. *Working software is the primary measure of progress.*
8. *Agile processes promote sustainable development. The sponsors, developers, and users should be able to maintain a constant pace indefinitely.*
9. *Continuous attention to technical excellence and good design enhances agility.*
10. *Simplicity – the art of maximizing the amount of work not done – is essential.*
11. *The best architectures, requirements, and designs emerge from self-organizing teams.*
12. *At regular intervals, the team reflects on how to become more effective, then tunes and adjusts its behavior accordingly.*

Summary

Agile product development goes far beyond the idea of a new project management method. Agile product development aims to change the existing organization and make it more competitive. The challenge is that agility does not work within existing structures, thought models, and systems – it claims to work on these aspects and change them. Agile product development aims to remove the obstacles that have been placed in the way of product development during the last 100 years.

This is based on three findings that question traditional process models and cultures:

- It is the developers who create value in a development organization. Supporting processes such as quality, purchasing, and management must be subordinate to the development process in order to make the organization successful.
- Product development cannot be planned to the same extent as production processes. Embracing variability in development and accepting transparency and truths is the basis of this new philosophy.
- The strength and innovations that set a company apart from the competition comes from the people, not from the processes. Processes must therefore be geared towards people, not the other way around. The continuous improvement of processes is a key competitive advantage.

Many of these ways of thinking emerged in the 1950s in the production at Toyota. Matthew May, an experienced trainer at the Toyota Academy, summarized the similarities (Mezick, 2012):

"I don't care what you call it ... whatever the technique is called, if it's a good one, comes down to two things: A commitment to respect people and a commitment to constantly improve."

People are already doing their best; the
problems are with the system.
Only management can change
the system.
W. Edwards Deming

Production and Lean

Why production?

Many ideas that are reflected in today's agile approaches have their origins in the principle of flow optimization in production. In this chapter, I will look at a few important aspects of production optimization and the topic of lean development.

Traditionally, production is controlled centrally. The planning of work processes then includes both the material flow and the work plans for individual production steps. Even if the processing times are relatively constant, in practice there are often capacity fluctuations, for example due to machine breakdowns, illness, special tasks. As a result, such central production planning and control systems (PPS systems) must be constantly measured and regulated in order to keep production running. This is normally only possible with the appropriate use of IT. Traditional manufacturing relies on increasing local efficiency through large batch sizes and high capacity utilization of people and production facilities. This contrasts with the concepts I present on the following pages.

Rethinking production

Toyota Produktionssystem (TPS)

At the end of the 1940s, Taiichi Ohno began optimizing vehicle production at Toyota. He was driven by the poor resource situation in Japan during and after the Second World War (Ohno, 2013). The aim was to produce the wide range of variants prevalent in Japan in small quantities in a cost-efficient manner. His means of achieving this was to reduce waste in production by increasing the quality of the process so that less waste was produced during the final inspection. In addition to rejects, Ohno identified the creation of buffers as another type of waste. Buffers are used to compensate for fluctuations in the process caused by quality problems and incorrect planning. Ohno began by solving the problems that had led to the creation of buffers in order to minimize or eliminate them altogether. The key idea was to always deliver the material to the production line at the right time, thus eliminating the need for intermediate stock. Today, we know this concept as Just in Time (JIT). The key to JIT production was a form of production control that started with the last process step and requested material from the upstream steps: Kanban, known as a pull system.

For me, the following aspect is important: Ohno changed how production was seen from local optimization of the individual process steps to an end-to-end process. Semi-finished products cause costs due to the capital value of inventories and warehousing, while at the same time slowing down throughput in production. The long lead time of large batch sizes also means that money is only made later. Clogged production lines are therefore an economically relevant factor. At Toyota, reducing the batch size to one piece flow alone reduced the cycle time by 90 per cent. In order for these smaller batch sizes to remain economical, Ohno had to make great efforts to reduce set-up costs.

In addition to reducing batch sizes, a key element for Ohno was to generate quality within the process and not just *add* it at the end of the process. In the new end-to-end approach, it was more beneficial for the company to stop production in the event of quality problems in order to solve the problem immediately and sustainably, instead of focusing on high capacity utilization and reworking the problematic vehicles at the end of the line. Another new feature was that the worker on the line could now decide to stop the production line without having to consult his foreman or another management body. For me, one of the key messages of TPS is that the costs that are saved by reducing inventory and avoiding waste are generally significantly higher than the costs that come about, for example, from lower capacity utilization when a production line is stopped.

A process chain that is efficient from end to end is always based on local inefficiencies in the individual process steps.

The organization, i.e., employees and managers, must be able to deal with the fact that it is perfectly fine if an employee has nothing to do or a production machine comes to a standstill. Priority is always given to immediately solving any problems that are encountered and maintaining throughput times while keeping quality at the same level. In the 1980s, the principles were taken up by James Womack and others and have since entered the language under the term *Lean* or *Lean Production* due to the streamlining of inventories.

Kanban in the produktion

When Taiichi Ohno was looking for a control option for his just-in-time concept, he was inspired by the logistics of American supermarkets. At that time, supermarkets did not order goods on the basis of central planning and estimates, but on the basis of actual consumption. If the stock fell below a certain level, the corresponding product was ordered. I deliberately write "back then" because today consumption is predicted in part using statistical methods based on available sales and environmental data and logistics are managed accordingly – and with surprising precision.

In principle, the shelves are replenished using data from the electronic tills, but even today you can discover a physical Kanban system in some stores: A card appears in front of the third-to-last product on the shelf, signaling the staff to reorder this product. This brings me to the literal meaning of the Japanese word kanban, which in this context is translated as card or more loosely as signal card.

While in centrally controlled production, as described, parts were produced in stock in the search for maximum efficiency, a production station in the Kanban system may only produce parts if it has the corresponding permission from the subsequent process step in the form of a Kanban card or of an empty Kanban container. This means – as you already know – that machine downtime is accepted in order to keep stocks low and maintain the flow in production.

With Kanban, each process step requests a production activity from its predecessor on an as-needed basis. This control runs from step to step through the entire production process (Figure 4). In this sense, the material to be processed is pulled through the process by the last process step, i.e., by the customer. This is why Kanban is also referred to as a pull system. In contrast, a central production control system pushes the material from the raw material warehouse into production, making it a push system. When comparing push and pull, please note that both systems start at the exact opposite end of

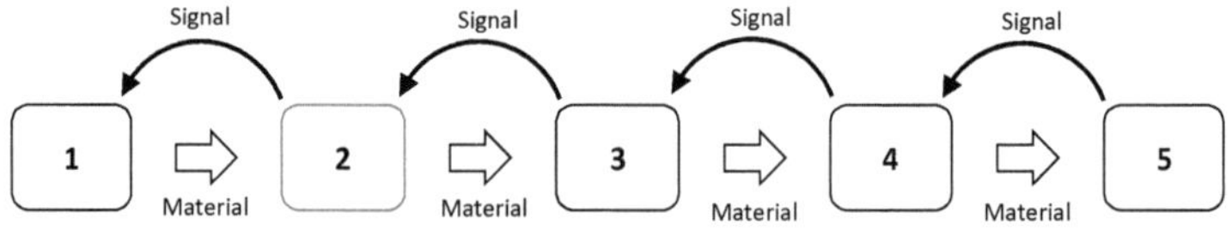

Figure 4: Kanban in production

the process chain. Push controls from the start of the process, pull from the end of the process.

At this point, I would like to pre-empt a discussion: Push systems can be just as effective as pull systems, they can also be run with small inventories. However, there is one major difference: For a push system to reach the level of a pull system in terms of throughput, a considerable amount of effort must be put into measuring and controlling the process chain. Pull systems, on the other hand, automatically regulate themselves to the optimum level targeted by their design.

Theory of Constraints and Drum-Buffer-Rope

Independently of the approaches developed at Toyota, another pull system in production emerged decades later: The Drum-Buffer-Rope system (DBR) that was developed from the Theory of Constraints. This was initially the basis for the use of Kanban in development, which I will present in a later chapter.

In the 1970s and 1980s, the Israeli physicist Eliyahu Goldratt developed several thought models to describe his understanding of processes in companies and summarized them as the Theory of Constraints (TOC). Goldratt's advantage was that, as a physicist, he had a very analytical way of thinking, but no experience in industrial companies. This enabled him to develop a neutral view of the typical problems in production processes without being preloaded with mental models. The Theory of Constraints came about by chance when Goldratt was helping a friend to optimize a chicken cage production process (Techt, 2010).

The core of this theory is production control that is oriented towards the bottleneck in the process chain and uses this as a control instrument. To explain his model, Goldratt chose the story of the scouts' hike, which I will briefly reproduce here (Goldratt & Cox, 1984): A group of boy scouts were assigned to march from A to B.

The route is a narrow path on which it is not possible to overtake. The goal is only reached when all the scouts have crossed the finish line. How does this analogy relate to the production process? An invoice can only be written once all process steps for the order have been completed. In addition, the group should block as little space as possible on the narrow path, i.e., use as few resources as possible.

Figure 5: Hike of the boy scouts

When the scouts' hike begins, the group looks something like Figure 5 after a short time: At the front, the group is spread out because every scout has their own pace. From a certain point on in the group, all the scouts walk closely together. The reason is that the scout at the front of the congested group is the slowest participant in the entire group. The scouts who could walk faster are right behind him, but they cannot go any faster due to a lack of overtaking opportunities. Goldratt named the slowest scout Herbie, which I will also use in my example. If your name is Herbie, please bear with me because by the end of the story, Herbie will be the most important scout in the group.

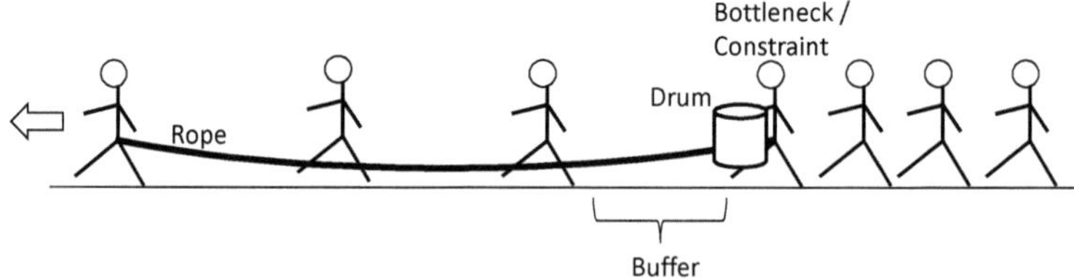

Figure 6: Boy scouts with Drum-Buffer-Rope

To get the group across the finish line as quickly as possible and with as few resources as possible, Goldratt devised the following control system (Figure 6):

1. The group must adapt to Herbie's pace. Herbie is therefore given a drum, which he has to beat at his own pace. Everyone else has to walk to the beat of this drum. The scouts with the longer legs are still a little faster.

2. A certain amount of space must be left in front of Herbie, a buffer must be created. Note: If Herbie stops, the group cannot make up for this interruption as Herbie is the slowest of them all. If the scout in front of Herbie now ties his shoelace, the buffer must be large enough so that the scout has finished to tie his lace before Herbie caught up with him.

3. It's not just the scout in front of Herbie who can suddenly have an a shoelace situation, Herbie could also suffer the same fate. If Herbie has to tie his shoelace, this time is lost for the group anyway. However, as the goal is only reached when everyone has crossed the finish line, and the path should be used sparingly, it makes no sense for everyone to continue ahead of Herbie if he interrupts them and the group is pulled apart as a result. To prevent this, Goldratt uses a rope and ties one end around Herbie's waist and the other end around the waist of the first scout in the group. If Herbie has to stop, the whole group stops.

I think this image of the scouts is a good way of understanding the bottleneck mindset. However, the implementation in production is not necessarily clear from this, so I would like to briefly discuss the Drum-Buffer-Rope concept in production.

If production is to be controlled with DBR, the bottleneck must first be identified, which can often be recognized by the visibly piled-up material in front of the bottleneck. As with the scouts, control should start from the bottleneck. In practice, the sequence of orders is planned at the bottleneck, which corresponds to the drum. All steps before and after the bottleneck work without further planning, according to the first-in-first-out (FIFO) principle. A buffer of material is built up in front of the bottleneck so that the bottleneck is always supplied, even if the process steps in front of it have problems with its throughput. To prevent production from filling up in the old push manner when the bottleneck has a problem, a new material feed into the production process is only released when the bottleneck has completed a job. This is the rope (Figure 7).

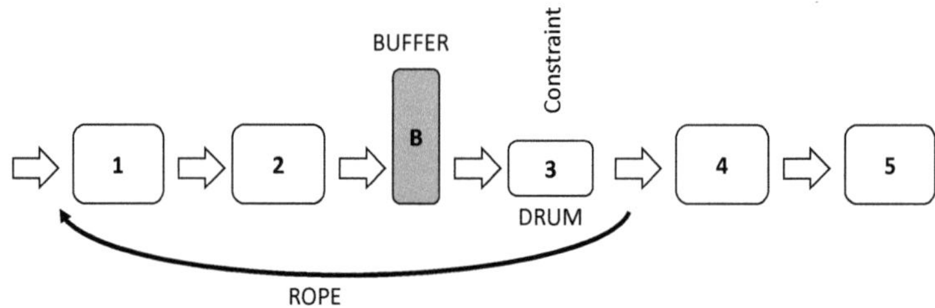

Figure 7: Drum-Buffer-Rope in production

In practice, the buffer is not defined by quantities, but by the lead time of the material feed. That means if order x is currently being processed at the bottleneck, order x+n (for n>1) is fed into the process when the current work is completed at the bottleneck. The number n automatically results in a certain buffer size. This blocks the feeding of further material if the bottleneck has problems with throughput. Drum-Buffer-Rope (DBR) is therefore also a pull system.

Important concepts and terms

Work in Process (WIP)

The two pull systems Drum-Buffer-Rope and Kanban only pull material into the system if it can also be processed promptly. I would like to introduce the term Work in Process (WIP) because we will encounter WIP continuously in the agile world. Note: WIP is often explained as *work in progress*. In my view, this is not entirely accurate. I prefer *work in process*. It is about how much work is currently in the system, i.e., in the process, and not what is currently being worked on. Semi-finished products are part of the WIP, but are not currently in progress.

Pull systems automatically limit the WIP without external control. This has several advantages in production: Inventory levels in production are reduced, which saves costs; and throughput times are shorter, which enables an earlier cash flow. Both effects are a consequence of the continuous material flow in pull systems. The flow is generally more continuous with Kanban than with drum-buffer-rope. The latter has the disadvantage that the filling and emptying of the system always takes place with a certain delay due to the rope and buffer, which in extreme cases can cause the pull control to swing up and down between the two: *completely full and completely empty* (Reinertsen, 2009). Kanban systems therefore fill up again more quickly after disruptions. Nevertheless, DBR is a proven and simple pull system in manufacturing that is much more widespread in the USA than in Europe.

WIP limits not only increase throughput, they also make systemic problems transparent. With a high WIP, the organization can work past disruptions or problems – after all, there is enough work in the process. With limited WIP, a problem could lead to a blockage of the entire system. This means that problems are detected quickly and can be tackled immediately, just like at Toyota in the production.

This constantly delivers small improvements, triggered by the blockages in the pull system. My favorite slogan: The blockage is a gold nugget for the process improver.

WIP limitation has a significant advantage in development and project management: It reduces harmful context changes among the people involved. You can find out more about this in the People and Teams chapter.

Lead Time and Cycle Time

I would like to briefly explain two important terms that are essential for understanding the lean approach. These are *lead time* and *cycle time*. Lead time is the time from when a task is entered into a system until it is completed, while cycle time is the time a task takes in the process. Figure 8 illustrates the difference.

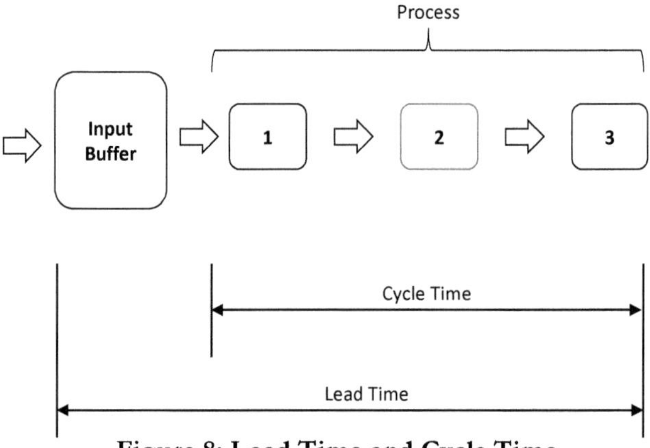

Figure 8: Lead Time and Cycle Time

The difference between lead and cycle time is therefore the waiting time of the tasks in the input buffer. In the case of uncontrolled processes without an input buffer, all tasks land directly in the process: in this case, both times are identical. As previously defined, the term WIP refers to the tasks in the process.

The aim of all lean/agile approaches is to introduce an input buffer and significantly reduce the cycle time by working on fewer things at the same time. The delivery time across all projects initially changes little as a result. In practice, however, the demand on a system in product development is always greater than its capacity. The aim must therefore be to consciously prioritize the important and urgent things first in the system and get them delivered within the shortest possible time.

Batch sizes

A batch is a unit of identical products that are controlled together through the production process. In contrast to one-off production in a factory-like environment, batch-based production offers major cost benefits. The larger the batch size, the less frequently the machines have to be retooled between orders — hundreds or thousands of parts are produced with one machine setting. These so-called set-up

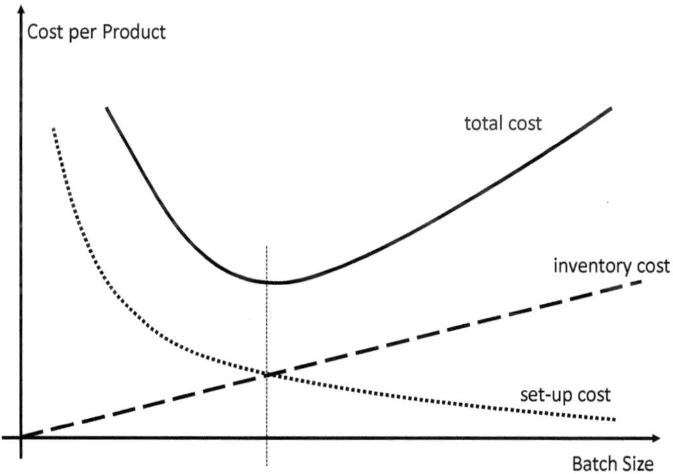

Figure 9: Cost depending on batch size

costs must be allocated to the individual production piece, resulting in a qualitative curve as shown in Figure 9.

Due to the focus on capacity utilization and production costs in the age of mass production, the disadvantages of large batches were initially overlooked. Large batches make production inflexible if orders are to be reprioritized in the interests of the customer. To a certain extent, large batches block production, which, in addition to the lack of flexibility, also entails high costs due to the material in production.

The raw materials and purchased parts have to be pre-financed on a large scale, and revenues are generated relatively late due to the long lead time of large batches. Overall, the sum of set-up and capital costs results in a curve of total costs, at the apex of which lies the optimum cost. This allows to determine the optimum batch size.

The difference between production batches and transport batches is important to understand. In production, large batches are not always transported between two work steps while waiting for the entire batch. Individual parts or transport containers are transported onwards before the corresponding work step has processed the entire batch. Transport lots are therefore often smaller than production batches.

Lean Development

In addition to the Toyota production system, the approaches in product development also attracted the attention of American consultants and authors: How did Toyota manage to bring new vehicles to market twice as fast as the US manufacturers — while at the same time achieving outstanding quality? This is the world of Lean Development. I would like to briefly introduce this way of thinking.

Waste in product development

The reduction or avoidance of waste is an essential approach in lean production. Types of waste such as rework, unnecessary routes or overload can easily be transferred to work in product development. In addition, Allen Ward introduced types of waste that relate to the fundamentally different work involved in product development (Ward, 2012). These are

- Scatter: This kind of waste distracts from the essentials. It arises, for example, through unnecessary reporting, overwork, constant queries, and organizational changes.

- Communication barriers: Typical barriers that cause friction and worsen quality include distributed teams, social barriers such as hard role definitions, unreachable managers or not taking lower-ranking colleagues seriously. However, there can also be barriers at a technical level, such as a lack of interfaces.

- Poor tools: This is not just about software tools, but more generally about processes or work instructions that unnecessarily slow down work without improving quality.

- Handoffs: Handoffs result from distributed responsibilities and accountabilities. This causes waiting times, and focus and a sense of responsibility also diminish.

- Useless information: Similar to the overproduction waste type, this is about unnecessarily generated information, such as management reports, status meetings, and so forth.
- Waiting: Development projects consist to a large extent of waiting, even if it doesn't feel like it due to multitasking: waiting for experts, for managers, for approvals, and for quality gates. The focus is often on capacity utilization and not on throughput.
- Wishful thinking: Wishful thinking arises when decisions are made on the basis of assumptions. For example, technical concepts are adopted without checking their feasibility. In development processes based on phases, wishful thinking is already built in.
- Testing to specification: Under time and cost pressure, tests are only designed to meet the specification, but not to find the limits of the product. This deliberately avoids learning more about the technical possibilities. In subsequent projects, this lack of knowledge has to be reacquired at great expense.
- Discarded knowledge: Organizations often place too little focus on learning, so knowledge gained can quickly disappear. For example, lessons learned are not implemented, no time is made available for documentation or knowledge about the organization's working methods is lost through restructuring.

Perhaps you have already identified your own organization in this list! If so, it's worth delving further into the subject as Toyota has a few decades' head start.

Shared responsibility

According to Allen Ward, handoffs are the most significant type of waste. What is it all about? In many organizations, there are precise role descriptions that define tasks and responsibilities. Product managers define the requirements for the product, system architects provide the technical implementation, development managers provide the developers and quality assurance is meant to ensure quality. This results in a large number of handoffs and very localized responsibilities. The problem is sometimes made worse by target agreements for different roles that sometimes contradict each other. This becomes most obvious when costs are to be reduced, but higher quality is required at the same time.

Lean Development's answer to this problem is to combine responsibilities. At Toyota, for example, there is a Chief Engineer who is project manager, chief architect, customer representative and engineering consultant at the same time. This brings together these four responsibilities, which are usually separate (according to Ward):

- Responsibility for the product & project
- Knowledge of the technology
- Knowledge of the market and customers
- Operational implementation

In practice, this enables the Chief Engineer to make entrepreneurial decisions such as adjusting requirements to reduce costs, or increasing quality by selecting suppliers according to criteria other than price aiding him in making risk-based decisions.

This approach is often not possible at project level, as it has a deep impact on the culture of the organization. However, it is an important key to developing better products in less time.

Build up knowledge

Building up knowledge is a central goal in Lean Development. Merely developing a product and transferring it to production is wasteful from this perspective – if the knowledge in product development is not significantly increased in the process. As described in the types of waste, it is therefore not enough to simply test for a given specification. The product must be tested to failure (statically or dynamically) in order to get to know the limits of feasibility. It is even better to build and test the product in several variants, especially if design parameters have to be weighed up against one another (e.g., weight vs. strength). Although today this can be estimated using simulations, prototypes in variants with correspondingly extensive tests are an important source of knowledge.

The knowledge gained can be documented in tradeoff curves. These curves show the limits of feasibility in a specific parameter space. At Toyota, thousands of such curves have been documented from tests for all possible components and their design and interaction. This means that even inexperienced developers could theoretically design a car or its components without having to make the experience themselves or performing special calculations.

Preserve options

Keeping options open goes hand in hand with acquiring knowledge. In particular, this involves pursuing several technical concepts (e.g. designs) in parallel until feasibility is confirmed through testing. All too often, phase-controlled processes force developers to make early design decisions, which entails a correspondingly high risk in terms of feasibility. In lean development, this is referred to as set-based design. However, keeping options open also occurs at other levels. Within the concepts, it is a good approach to use value ranges instead of concrete values in the specifications. This gives the development team more freedom to find the best economic or technical solution. Fixed specifications prevent opportunities. And at higher levels, for example in the project portfolio, it can make sense to pursue different strategies at the same time if it is not clear how markets and framework conditions will develop (e.g., parallel development of battery-electric and hydrogen-powered vehicles).

Lean Development: Conclusion

Unfortunately, it would go beyond the scope of this book to delve deeper into these topics. I hope I have been able to arouse your curiosity with the selected aspects of lean development. If you would like to read more, I recommend *Designing the Future* (Morgan & Liker, 2018) to get you started.

Lean Development – second generation

Donald Reinertsen, an expert in product development, has founded a movement that goes beyond classic Lean Development and is therefore often referred to as the second generation. Reinertsen has been studying the nature of development projects for more than 30 years and has contributed significantly to a deeper, alternative understanding of interrelationships in development. I personally consider his book *The Principles of Product Development Flow* (Reinertsen, 2009) to be the most important book on the nature and optimization of product development.

Variability in production vs. development

It makes sense to apply concepts from the Toyota Production System or lean manufacturing to product development. While the solution approaches can be easily transferred from production to development, there is also a risk that the same goals will be pursued as in production, although the focus should be on completely different goals. Product development differs from production in one key respect: While the aim in production is to keep the variability in the process low in order to achieve reproducible results, it is in the nature of development not to be reproducible. Every single time something completely new and unprecedented comes about. Variability in development processes is therefore at the heart of what we call innovation. Anyone wanting to make development lean without reflection through the production-coloured glasses usually kills what development is all about. Instead, we need to use the lean approach to reduce lead times and to better deal with the existing variability.

Costs and opportunities in development

Anyone who develops products and launches them on the market or uses them for themselves does so for one or more of the following four reasons (Arnold & Yüce, 2013):

1. Increase in revenue: The product is to be sold on the market.
2. Protection of revenue: The product is intended to protect existing sales against competition.
3. Cost reduction: The product simplifies processes.
4. Cost avoidance: The product eliminates a certain type of cost.

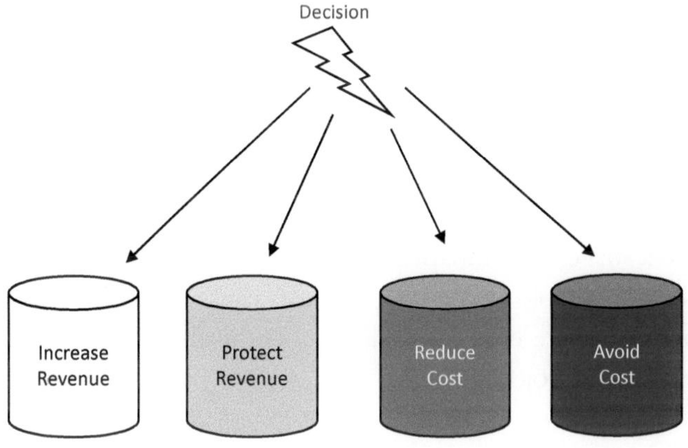

Figure 10: Four impacts according to Arnold

These four aspects also apply to every decision made in a development project. Everything you do will contribute to at least one of these four aspects, see Figure 10. In practice, however, these aspects are not as independent as depicted. As an example, let's look at the relationship between increasing revenue and reducing costs. If you reduce costs in development and these costs do not represent waste in the lean sense, you will almost certainly lose revenue because the development project is delayed or the developers are forced to cut back on quality or maintainability.

In many organizations, this relationship is neither known nor quantified. Different people or departments are usually responsible for costs and opportunities, for example product management for the opportunities and controlling for the costs. There is rarely an exchange or even agreement on what cost savings in development subsequently destroy market opportunities. Another contributing factor is that those responsible for opportunities think and plan over long periods of time – the product life cycle – while those responsible for costs think and plan in terms of annual budgets. The decision to cut costs in the development project, for example by withdrawing resources or not procuring the necessary infrastructure, destroys market profitability, which usually amounts to a multiple of savings. If a holistic view is taken, the opposite decision would often have to be made. However, this consideration rarely takes place. Donald Reinertsen surveyed companies and came to the conclusion that only 15 per cent were able to quantify the costs caused by delays. The other 85 per cent were unable to make economically sensible decisions in development projects.

Cost of Delay

Many development organizations are still geared towards maximum capacity utilization. Many understand the high efficiency required everywhere to mean high-capacity utilization. This is understandable, as it is easy to calculate what it costs the organization if an employee is underutilized for an hour. I'm sure you also have reliable data on what a developer, including workplace and infrastructure, costs every month or day. As far as the goal of high-capacity utilization is concerned, the parallels with manufacturing cannot be denied. If the production world came to the realization 70 years ago that too much inventory causes more costs than too little efficiency, how can this now be applied to development projects?

Why not conduct a short survey among your colleagues, employees, managers, and project leaders? Ask everyone individually what costs the company incurs if a project is completed one day later than planned. If your projects are linked to a fixed deadline and a possible contractual penalty, such as in the automotive industry, turn the question around: What is the profit to the company if a milestone is completed a day early?

Donald Reinertsen has often asked his customers these questions and found that the respondents either had no answer at all or that the estimates of the individual project members differed by a factor of up to 50. As described above, only a few organizations can give a valid answer to the question of Cost of Delay (COD). It is probably the case, and you may be able to confirm this from your own experience, that most organizations are not aware of what it costs not to complete a certain task for a day. As confirmation, I recommend taking a look at the vast number of tasks on open points lists (OPL) or in an issue tracker in software development.

The problem in development is that in contrast to the material in production, tasks in development are not visible. They are hidden in heads and on hard disks. If you succeed in quantifying your Cost of Delay, you have a very powerful tool in your hands. If you can quantify COD, you can justify many decisions that were previously made on instinct from a business perspective: Is it worth putting more resources into the project? Is it worth purchasing a new tool for development? What is the ideal capacity utilization? What is the ideal lead time?

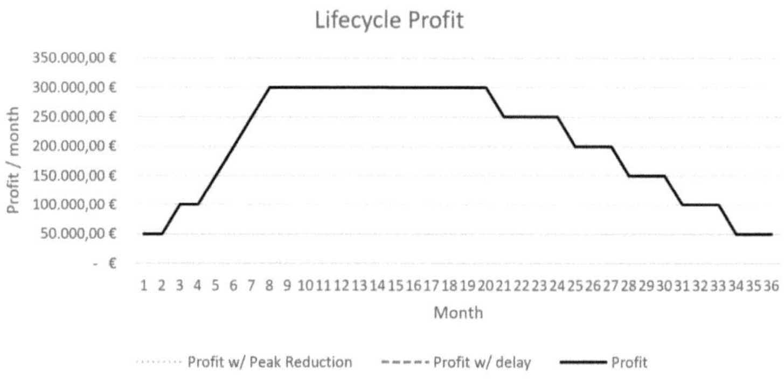

Figure 11: Profit (estimated)

At this point, I would like to use an example to illustrate this. Let's take a mechanical engineering company and its (fictitious) product management sales and profit forecast. As you can see, the product life cycle covers approximately three years. The new machine is launched on the market and then sold over a period with six units sold per month. The profit per machine is assumed to be EUR 50,000. Subsequently, the sales figures fall again due to market conditions (Figure 11).

So, what happens if the development project is completed one or two months later? The frequently encountered misconception is that this curve is then simply shifted to the right by one or two months, i.e., that the profit achieved over the product lifetime (lifecycle profit) remains constant and only the costs for additional months of development worsen the calculation.

The problem is that as a company, you can only determine the ramp-up of the product launch. You have no control over when the product is taken off the market. This is determined by the market and the competition. However, if the market launch is delayed, your lifecycle profit curve shortens. All the area you lose under the curve is lost profit. There is also a further effect: If you enter the market later, your market share may decrease. For the curve as demonstrated, this means that it does not rise as high as initially assumed.

If we assume a delay of one month in our example and a reduction in the maximum by only five per cent, the new curve looks like the one shown in Figure 12. Graphically, the lost areas under the curve do not look that significant. Expressed in figures, however, they are: In this project, you lose EUR 650,000 with a delay of one month, which is more than EUR 21,000 per day and around EUR 900 per hour.

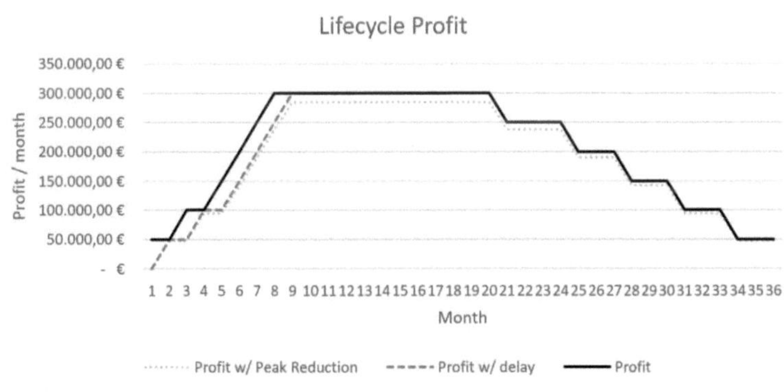

Figure 12: Profit (delayed)

An unavailable tool, an absent employee, a postponed management decision, a slow procurement process for prototypes can really ruin a business economist's day without it showing up anywhere in the existing figures.

So where do delays occur in product development? I will present some aspects of this in the following sections.

Practical tip: Calculation of COD

In contrast to costs, which are usually known in detail, the figures behind the calculation of COD are often only estimates and therefore inaccurate or even contestable.

I believe that for the time being it is absolutely sufficient to determine the rough dimension of COD. The figures that come out of the initial calculations will have a significant impact on the organization's thinking, even if they are off by a certain factor.

Even a very rough estimate of COD is much more useful than having no idea at all.

Queues

Donald Reinertsen has been using queueing theory to explain mechanisms and effects in product development for many years. The basis for this is that the tasks that come to a development organization have broad statistical distributions – both in terms of the time of arrival and the effort involved in the tasks. This fundamentally distinguishes product development from production, in which the processing times of the individual process steps have a very narrow spread. Incidentally, this is also the reason why classic planning attempts based on assumptions from the world of production regularly fail in product development. But that is precisely why you are reading this book.

In his search for industries that were used to dealing with broad probability distributions for requests, Reinertsen came across the telecommunications industry, or to be more precise, the queuing theory. The Queuing Theory, developed in 1909 by mathematician Erlang for telephone companies, relates the utilization of a server to the length of a queue that forms in front of the server. The term server is not to be seen in a technical sense here; a server simply processes tasks. In our context, this is, for example, a manager, a technical expert, a development team or a supplying department. For the binomially distributed request frequencies and task sizes to be assumed in development, the formula in Figure 13 results is for a setup with one queue and one server.

$$L = \frac{\rho^2}{(1 - \rho)}$$

L = Lenght of the queue (queue size)

ρ = Utilization of the server (rho)

Figure 13: Formula for queue size

Graphically, you can see at a glance the relationship between the utilization (efficiency) of the server and the length of the queues, see Figure 14.

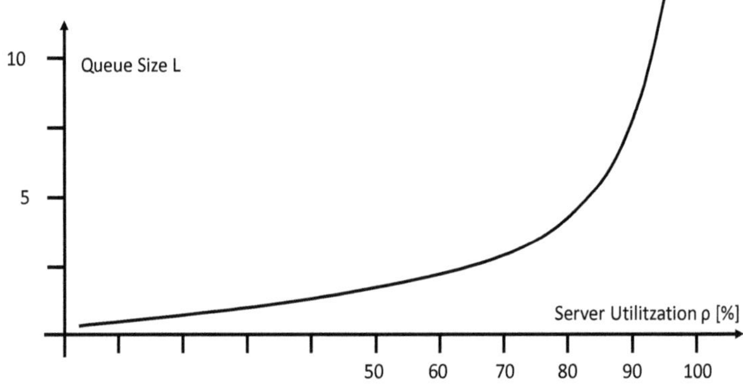

Figure 14: Queue size

It is important to know that the lead time of a server is linear to the length of the queue. You can see four important things in this diagram:

1. The relationship between capacity utilization and lead time is not linear. From a capacity utilization of 80 or 85 per cent, the curve becomes very steep and the lead time increases dramatically with increasing capacity utilization.
2. On the other hand, with high-capacity utilization, even a small reduction in workload results in a significantly shorter lead time.
3. The effects of estimation errors can be explained well in the diagram. An estimation error in the sense of "underestimating" increases capacity utilization. If your organization is currently operating at 70 per cent capacity, a five per cent increase will result in a 38 per cent longer lead time. If your organization

Production and Lean

runs on a capacity utilization of 90 per cent, an increase of five per cent already pushes up the delivery time by 123 per cent.

4. As the utilization of a person, team or department is very difficult to measure in practice, the curve can be read back to front: Based on the length of the queue, which is easy to measure, you can determine the utilization.

Even without using the queueing theory quantitatively in your company, you can see from the diagram that neither zero per cent capacity nor 100 per cent capacity is a desirable operating point for your organization. While this is still logical at zero per cent (nobody is working any more), the 100 per cent point does give cause for thought: At maximum efficiency of your employees, the throughput times in the development process approach infinity! Optimal efficiency is therefore not the same as maximum efficiency.

It is important to understand that the relationships I have described so far in queueing theory describe static relationships. The formulas describe average values for the steady state. This is important in order to internalize the basic relationships. However, the dynamic aspects of queues are interesting for day-to-day work: Queues are very early indicators of capacity utilization and lead times. By observing the length of queues, it is possible to react to developments much earlier than with the traditional measurement of delivery times. You know this approach from supermarkets, for example, supermarket staff decide when to open another checkout based on the length of the current queue at one. Nobody measures waiting times in the supermarket or even plans when customers should get to which checkout.

Practical tip: Identify queues

The first thing you should do is develop an awareness of queues in your organization by identifying servers and visualizing queues, either by using sticky notes or paper lists. This will give you a quick overview of workloads and delivery times. If you already have initial estimates of COD, you can also calculate the costs of the queues. This alone will influence many discussions and decisions.

According to Donald Reinertsen, queues have preferred places in organizations:

- Marketing
- Experts
- Engineering
- Purchasing
- Prototype construction
- Test and test stands
- Management Reviews
- Toolmaking
- Other technical experts

It is worth looking, measuring, calculating and optimizing in this case. Agile working methods such as Scrum or Kanban offer support in dealing with queues and the opportunity for gradual improvement and change.

Development batches

However, the Costs of Delay described above do not only occur in queues caused by high-capacity utilization. Queues also arise at points in the company where tasks have to wait for other tasks for process-related reasons because they are processed in batches. Both types of queues generate inventories in development, which – according to the considerations on COD described above – cause significant costs.

Earlier, I described how Taiichi Ohno at Toyota was able to reduce lead time by 90 per cent simply by reducing the batch size. Reducing the batch size is therefore essential in order to reduce throughput time and inventories and therefore cutting costs. The fact that batches exist in development is not obvious at first. The question "Where are tasks waiting for other tasks?" helps in the search for batches. Stage-gated processes, in which the project status is switched from one project phase to the next according to certain criteria, are an extreme example with regard to batch sizes. In such processes, the maximum possible batch size is used, which also maximizes costs according to the batch size diagram. The situation is similar when it comes to reviews: several documents often go into reviews together, which means that many documents will cause COD while they wait for the last document in the review batch.

Less obvious than the process-related batches just described are batch sizes that arise from requirements that are too large or from projects that are too big. The latter results from the widespread budget culture with the effort required to get a project budget approved leads to projects being as large as possible. Several small projects in a row on the other hand, would reduce the batch sizes *and* the Cost of Delay.

However, it is not always the process and organization that drive batch sizes in development; it is often the products themselves. As soon as the product is no longer just a software product, but also includes mechanical and electronic components, batches inevitably arise as soon as the product is materialized. When building samples in electronics and mechanics, several requirements and parts must inevitably be combined.

A *one-piece flow* of requirements, as with software, is generally not possible in this case. Therefore, it also makes sense to combine several tests for cost reasons.

This intuitively understandable correlation can also be illustrated in business terms by showing the various cost types in relation to the batch size. You will be familiar with this relationship from the section on batch sizes in production. In development, Cost of Delay replaces inventory costs. Instead of setup costs, we are talking now about transaction costs, which are costs for design, construction, and for example, testing. The diagram remains the same while the inventory costs or COD increase linearly with the batch size, the transaction costs per unit increase exponentially as the batch size decreases. The sum of both costs then represents a trough-shaped curve, at the lowest point of which lies the optimum (Figure 15).

The key aspects in this context are batch sizes that are too small – and therefore lead time is no more the optimum than a batch size that is too large. Secondly, a reduction in transaction costs inevitably leads to a reduction in overall costs. Things like 3D printing, simulations, test automation and so on should be considered carefully from this perspective. The distinction between processing and transportation lots can also be applied to product development. For example, it is useful not to wait for a final test report for a large test that checks for many requirements and runs for a correspondingly long time, but to regularly feed interim results back to development.

At the end of this consideration, I would like to share a thought from Donald Reinertsen: The correlations taken from manufacturing above assume that transaction costs are independent of batch size. In fact, set-up costs in manufacturing, for example, *are* independent of the batch size. In development, on the other hand, especially when managing requirements, bugs or change requests, transaction costs may also depend on the batch size. This effect is caused by overhead costs when large amounts of data are handled by people. This makes the mechanisms described even clearer and the demand for smaller batch sizes even more urgent.

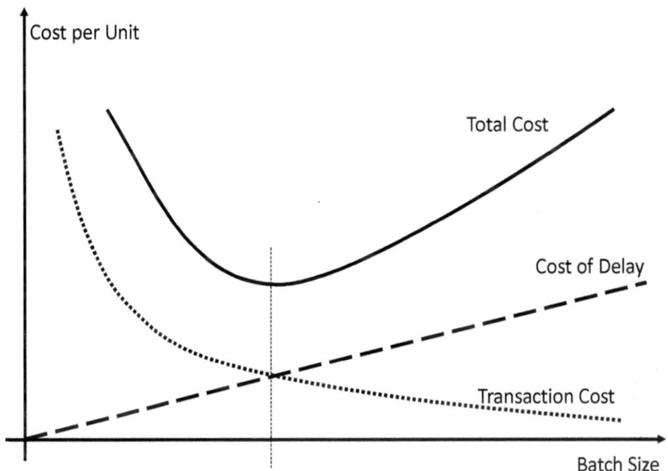

Figure 15: Batches in development

Decision paths

Votes, i.e., decisions in which several people are involved, can in principle be made in a synchronous or asynchronous way.
werden.

Asynchronous votes are votes by circulation. In practice, I have seen examples of this in large companies where document reviews by circulation take a duration of two to three months. If you like, you can calculate this using the example described above for COD of the fictitious mechanical engineering company. The runtime is made up of two unnecessary aspects: Firstly, the idle time on the reviewers' desks, and secondly, the multiple circulation of documents when comments have to be shared again with everyone else. Figure 16 is an example of the flow of information in asynchronous decisions.

Conducting votes with minimal COD means relying on synchronous voting: Everyone involved must be at the same table at the same

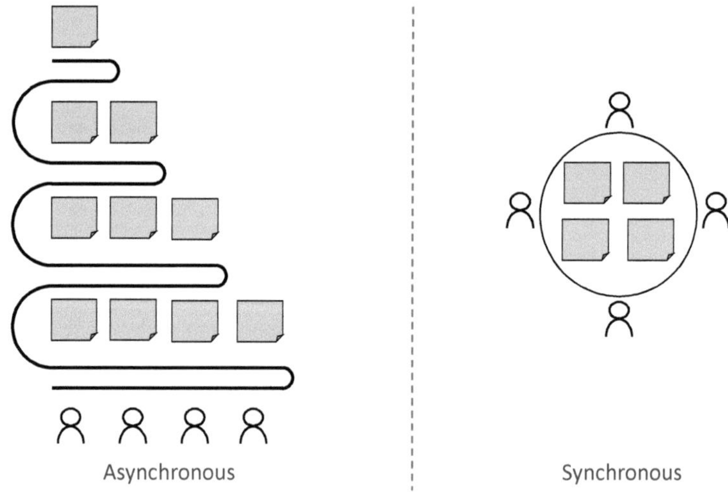

Figure 16: Asynchronous and synchronous decisions

time. My observation is that this is often unpopular because employees already suffer from a high number of meetings. As a result, synchronous coordination is often perceived as inefficient, although the exact opposite is the case. At the latest when you are seriously considering your COD, you have no choice but to switch to synchronous decision-making processes.

WIP Limits

When explaining the concepts of production, I already briefly touched on the importance of limiting concurrent work in process (WIP). First of all, limiting work in process leads to a reduction in throughput times. If you work on fewer tasks at the same time, you will finish the individual task faster. That is plausible. So, what is the point of setting up an input buffer and only adding tasks to the process in chunks, i.e., limiting WIP? There is an unexpected increase in speed due to three effects:

1. The change of context between tasks is reduced and so is the waste.
2. Significantly increased optimization potential for processes thanks to greater transparency.
3. Significantly shorter throughput times by focusing resources and reducing queues.

Context change is covered in the People and Teams chapter. Greater transparency arises because hidden problems in the organization or in the process become visible when a WIP limit creates a blockage in the flow. Without a WIP limit, it is tempting to work past problems instead of tackling and solving them. WIP limits therefore create the necessary pressure to work on the system and improve it instead of trying to somehow get through the day in the system.

The third effect in my list is the most important to me at this point. Even a very high, defensive WIP limit has a major impact on throughput. Donald Reinertsen has calculated the following figures for an organization with 90 per cent capacity utilization as an example: A WIP limit of twice the average WIP already shortens the cycle time by 28 per cent, yet the idle time only increases by one per cent and only one per cent of requests have to be put on hold. That means, if you have an average WIP of 10 tasks, you would put the 11th task on hold until another of the 10 tasks has been completed. A much more aggressive WIP limit of half the current WIP, in our example that is five tasks, would reduce the throughput time by 72 per cent, but at the cost of having to reject 13 per cent of incoming tasks and increasing the idle time by 21 per cent. These calculation examples illustrate the great leverage that WIP limits have – and the disadvantages associated with limits that are too aggressive. In practice, the ideal limit is usually determined empirically and adjusted again and again based on experience. More on this later in the chapter on Kanban in development. The easiest way to implement WIP limits is to use pull systems such as Drum-Buffer-Rope, Kanban, Scrum or a simple pull system around a team, a department or an infrastructure, as demonstrated in Figure 8.

Stop running the relay race
and take up rugby
Hirotaka Takeuchi & Ikujiro Nonaka

Scrum

History

First building blocks

I will shed some light on the history of Scrum from the perspective of Jeff Sutherland, one of the two inventors of Scrum. As I ran Scrum training courses with him from time to time, I know the development of Scrum mainly from his point of view and I am sharing a few of the anecdotes that describe his path to Scrum here.

Jeff Sutherland served as a jet pilot in the Vietnam War. Later, as an officer, he succeeded in sustainably improving the marching precision of his company during parades by writing down errors and problems on index cards, pinning them to the company board and leaving it up to the soldiers themselves to define measures and training. He observed how his company improved faster through this transparency than with the measures prescribed by his predecessors. Decades before Scrum became widespread, Sutherland had already found an essential building block: transparency through visualization and the resulting self-organization of teams.

In the 1980s – Sutherland had meanwhile completed his doctorate with the development of a simulation of cells and cancer cells – he experimented with various practices to make software projects more successful. An interesting episode in this context is the invention of the burndown charts. A project manager told Sutherland about his observation that almost all projects would overshoot their time target. He was therefore looking for a method that could be used to achieve precision landings in project management. Sutherland, having been a jet pilot, was familiar with the problem of overshooting when landing and knew the recipe for precision landings: a precise and smooth approach. So, he used what he knew from flying as the vertical profile of an approach as a burndown chart for projects. The ideal approach line is specified, with a target/actual comparison and small corrections to the flight altitude being made at fixed intervals. As a holder of a commercial pilot's license, and ever since I heard Jeff Sutherland tell the story of how they came about I now look at these charts differently.

Further findings and the first agile building blocks gradually came together to form Sutherland's picture of the agile implementation of software projects. This resulted in Scrum in the 1990s.

The birth of Scrum

In 1993, Sutherland continued to develop his approach at Easel Corporation, influenced by various studies and publications dealing with Lean, throughput optimization, team orientation, and so forth. The Toyota Production System (TPS) was decisive for Sutherland.

The term Scrum comes from a highly regarded paper by two Japanese scientists on the development of new products, the results of which were published in the Harvard Business Review in 1986 under the title *The New New Product Development Game* (Takeuchi & Nonaka, 1986). Hirotaka Takeuchi and Ikujiro Nonaka examined product development projects that had produced extraordinary innovations in a very short time. The products in the study: photocopiers, cameras and even a car – interestingly, not just software projects.

The following quote from the authors summarizes the quintessence: Stop running the relay race and take up rugby. They had observed that in these successful projects, the approach was not based on a division of labor and phase-controlled (e.g., relay race as in requirements, design, construction, and testing), but that all specialist disciplines worked as a team moving across the rugby pitch together. These cross-functional teams had a high degree of freedom in their working methods and took an iterative and incremental approach to development. A standard situation from rugby: a scrum (players packed closely together around the ball), became the eponym for the new methodology developed by Sutherland in this publication and later for the new way of thinking in product development.

In addition to Toyota and *The New Product Development Game*, the work of Sutherland and his colleagues at Easel was influenced by a publication about a development team at Borland, the software team with allegedly the highest productivity ever measured. It was in this environment that the idea of the daily coordination meeting was born.

In 1995, Ken Schwaber, then CEO of a project management software company, began to convert his development to Scrum and, together with Sutherland, to publish his experiences to date at conferences. The paper presented at the 1995 OOPSLA conference with titled *SCRUM Development Process* (Schwaber, 1995). From this point onwards, Scrum became known to the general public.

Scrum today

After the OOPSLA conference, other organizations began experimenting with Scrum. Sutherland and Schwaber now also received external feedback, from the newly emerging agile community. In 2002, they wrote the book *Agile Software Development with Scrum* with Mike Beedle. In 2001, these three authors were also co-creators and signatories of the aforementioned Agile Manifesto. Since 2011, Sutherland and Schwaber have maintained the Scrum Guide, the official definition of the Scrum framework. Even today, the Scrum Guide is regularly updated based on feedback from the agile community. The current Scrum Guide can be downloaded at www.scrumguides.org. Translations in many languages are also available on this website. Today, Scrum is used as a generic project management framework and has long since left its original domain, software development, which was illustrated even more clearly in the 2020 edition of the Scrum Guide. Mechatronic product development, administrative teams in purchasing or sales and management teams, including top management, have discovered the benefits for themselves: empirical planning and close tracking, the power of teamwork and, above all, the performance booster – focus.

An often-cited example is Saab's Gripen fighter plane, which was developed entirely using Scrum by over 1000 developers (Furuhjelm, Segertoft, Justice & Sutherland, 2017). In contrast to many conventionally organized projects of this size, this one was exactly on schedule and within budget.

Scrum Guide 2020

The authors have made some significant changes in the Scrum Guide 2020. If you are familiar with older versions, these points may be of interest to you:

- Substitution of the term Roles with that of Accountabilities, eliminating the term Development Team completely
- Elimination of all formulations related to product development, such as requirements, tests, etc. Scrum is now a framework for solving complex tasks.
- Introduction of the Product Goal as the objective of development.
- Positioning of Product Goal, Sprint Goal, and Definition of Done as *commitments* to the respective artifacts.
- Omission of some guidelines and recommendations (e.g., three questions in the Daily Scrum) to make the Scrum Guide more universal and clearer.

Scrum Framework

Framework or process?

Scrum is not a process description or work instruction, in fact not a method in the classic sense. Rather, Scrum is a framework that provides rules and a playing field. Within these guidelines, each team and each organization define the process that suits them and their product and constantly optimizes it.

These context-related improvements are only made possible by a framework concept. This is because firmly defined processes do not allow a team to find and follow the best and fastest way to realize the product.

Many things that are associated with Scrum – for example task boards, burndown charts, and user stories – are not part of the framework. Scrum is limited to the absolute minimum in order to make it work. You should take this into consideration in case you are tempted to leave out Scrum elements at a later stage. And that moment will almost certainly come.

Empirical process control

According to Dave Snowden's Cynefin model, an iterative action strategy is required in a complex environment. The prerequisite for iterative optimization and self-organization is maximum transparency of the current situation.

Accordingly, the Scrum Guide defines a loop consisting of Transparency, Inspection, and Adaptation as the basis for empirical process control (Figure 17). This triad of transparency, inspection and adaptation is often referred to in the agile community as *inspect and adapt*. This emphasizes the advantage of short control loops and experiments as opposed to long analysis and planning phases. In my opinion, however, it ignores the most important issue: Transparency.

However, transparency is not always desirable, as it requires a sense of safety on the part of everyone involved. In other words, there are no sanctions if problems become apparent. Only when systemic problems become visible is it possible to improve the organization in small steps and ultimately transform it into a learning organization.

My favorite phrase when starting agile training is, "Agility can't do magic. It will only make things transparent. And you won't like what you will discover." Transparency alone is therefore not enough. Only those who have the will and the authority to change things can use the power of transparency to benefit their organization.

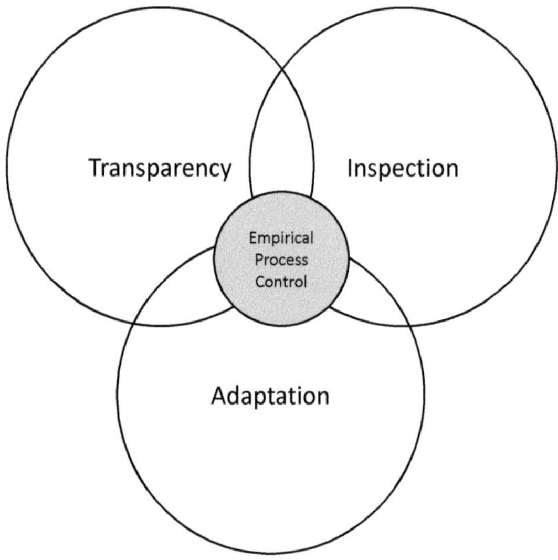

Figure 17: Empirical Process Control

Timeboxing

An important concept for agile working methods, especially in Scrum, is working with timeboxes or timeboxing. A timebox is a time limit for activities and meetings are stopped as soon as it expires. This makes it possible to plan your own actions and helps to make meetings more efficient. After a more or less painful learning curve, defined timeboxes will become part of everyday team life. This rule does not mean extending activities or meetings to the end of the timebox; they can of course also be ended earlier. The reference to the Scrum framework: Scrum sets upper limits for event timeboxes, the specific length is determined by the Scrum team.

Scrum at a glance

Scrum defines three accountabilities, three artifacts and five events as well as their interaction (Figure 18). These are

Accountabilities
- Product Owner
- Developer
- Scrum Master

Artifacts
- Product Backlog
- Sprint Backlog
- Product Increment

Events
- Sprint
- Sprint Planning
- Daily Scrum
- Sprint Review
- Sprint Retrospective

I will first introduce the Scrum way of working and then go into more detail about accountabilities, artifacts, and events in the following chapters.

Product development with Scrum is based on stakeholders with a Product Goal. This product development goal serves as a beacon throughout the entire development process. Incidentally, stakeholder is not a role according to the Scrum definition.

In discussions with stakeholders, and based on the Product Goal, the Product Owner draws up a list of all the things that need to be done in the course of implementing the product. This list is referred to as a Product Backlog. A backlog is a clearly prioritized list in which there are never two entries with the same priority, with the most important entry always at the top. The Product Backlog usually includes requirements or tasks, the Product Backlog items (PBI), which can also exist in different granularities. In a complex environment, it makes sense to define only the top entries in the

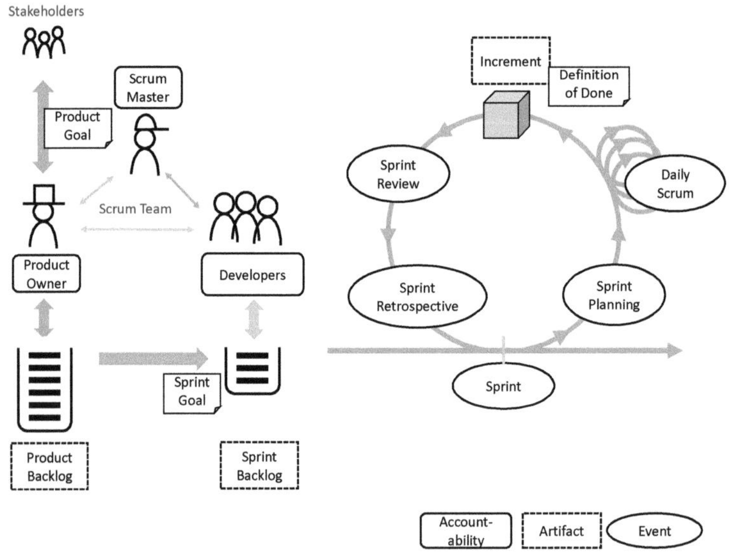

Figure 18: Scrum at a glance

Product Backlog precisely and to formulate the items further down more vaguely. The reason being is that the probability that Product Backlog Items further down the line will change is high. Early detailing would be a waste in terms of lean thinking.

Working in a Sprint

The Developers develop the product independently by working through the Product Backlog items in the order specified by the Product Owner. As with the Cynefin framework, in a complex environment the *probe, sense, respond* loop is run through again and again on the way to the goal. The Developers therefore do not work through the Product Backlog in one go, but work in short planning and development cycles of a few weeks, called Sprints. The Sprint is a container event that includes the other four events.

The scope of work for a Sprint is defined exclusively by the Developers, as they also carry out the work. Starting at the top, they take as many PBIs from the Product Backlog as – in their opinion – fit into the next Sprint. The Developers place these selected Product Backlog items in the Sprint Backlog, retaining the original prioritization specified by the Product Owner. While the developers work on the Sprint Backlog during the Sprint, the Product Owner can make changes to the Product Backlog at any time and reprioritize, change, add or remove items.

Planning

The Sprint scope is defined in Sprint Planning, the first event in the Sprint. In the Sprint Planning, the Product Owner communicates to the other team members what the next Product Backlog items are about and the Developers provide information on dependencies and effort. Furthermore, the Developers begin to break down the selected Product Backlog items into smaller tasks during Sprint Planning. In other words, they start with the concrete work planning for implementation. These tasks are then also part of the Sprint Backlog. They are not yet assigned to individual team members in Sprint Planning in order to remain flexible during the sprint.

On the road

After the Sprint Planning, the Developers start the development process. They meet every day during the Sprint for the Daily Scrum, where they take a brief look at the results of the last 24 hours and plan and agree on the steps for the next 24 hours.

Parallel to development work, Product Owners and developers continue to work on the Product Backlog. As with Sprint Planning, the aim is for the Developers to understand and estimate the items. In the course of this, larger or coarser items that have now moved further up the Product Backlog are defined more precisely and broken down into smaller packages. This procedure is called Product Backlog refinement, but is not an official Scrum event. The authors of the Scrum Guide give the teams the freedom to schedule the refinement as a meeting or to implement it as an ongoing activity.

End of the Sprint

By the end of the Sprint, the Developers have created a tested Increment, the third official Scrum artifact (Transparency), and present it to the Product Owner in the Sprint Review (Inspection). The Product Owner and, if necessary, the stakeholders evaluate the Increment. Change requests are added to the Product Backlog (Adaptation) and are prioritized by the Product Owner and adjusted to the existing Product Backlog items, as the Sprint is not extended in the event of change requests. The Sprint Review closes the feedback loop in terms of the product. Before the Sprint in this exemplary description comes to an end, I would like to briefly touch upon the previously unmentioned responsibility of the Scrum Master. He does not make any decisions regarding content and does not intervene directly in the development process. He helps with the implementation of Scrum by being available to the team as a moderator and mediator, and he is responsible for organizational issues and processes. He takes care of optimizing the team's working environment, pushes for changes in the organization and therefore giving the product development organization more time to remove obstacles, or impediments.

Sprint Retrospective

The last event within the Sprint is the Sprint Retrospective. This is where the Scrum Team, i.e., the Product Owner, Developers and, Scrum Master, discuss the progress of the current Sprint and look for ways to improve the way of working. The Sprint Retrospective closes the feedback loop regarding the working method or process. The end of the Retrospective also marks the end of the Sprint, and the new Sprint starts immediately thereafter.

What happens if the Developers have worked through all the items in the Sprint Backlog before the end of the Sprint? They then move onto the next item in the Product Backlog, but only if they are sure that this item can also be implemented in the remaining time. If items remain unprocessed in the Sprint, they are returned to the Product Backlog and the Product Owner prioritizes them with the existing items already there. In Scrum, the cadence of the Sprints is therefore maintained and the content of the Sprint is modified in the event of incalculables – these are usually estimation errors, technical problems, and disruptions.

After this overview, I will move onto the following subchapters in the detailed definition of the Scrum framework.

Accountabilities

Accountabilities replace Scrum roles in the Scrum Guide 2020. This is intended to further emphasize shared responsibility and the elimination of hierarchy. There is now only the Scrum Team and no longer, as previously, a Development Team within the Scrum Team.

Product Owner

The Product Owner is responsible for the What. They act as an interface to the stakeholders, and therefore to the market, and maintain the Product Backlog. Strictly speaking, they can also delegate the content-related work in the backlog to others, for example, the Developers or stakeholders. What definitely remains their responsibility, however, is the prioritization of the items in the Product Backlog. And it is the only person in the organization who decides on prioritization. That means that they are solely accountable for the costs, content and schedule of product development.

The Product Owner decides how and when to prioritize the Product Backlog. In practice, the business value of the requirements usually plays the main role with the aim to take the most valuable product functions to market as early as possible. Other important aspects in prioritization are external deadlines like trade fair presence and changes in legislation, the risk behind the requirement, but also quality aspects such as architecture adjustments, and redesigns.

The Product Owner is the contact person for the Developers when it comes to understanding tasks and requirements. They work closely with the Developers, particularly during Sprint Planning and backlog refinement. They should also be close to the team during the Sprint in order to be able to answer questions quickly. The Product Owner can therefore only achieve a good level of efficiency for his Scrum team if he has far-reaching decision-making powers and does not have to clarify every little thing with the stakeholders first.

The Product Backlog can be changed by the Product Owner at any time without affecting the work of the Developers, as they remove the PBIs for the Sprint from the Product Backlog and file them in the Sprint Backlog.

At the end of the sprint, in the Sprint Review, the Product Owner assesses the current Increment and, if necessary, submits change requests in the Product Backlog, which are then processed in one of the next Sprints. In order to obtain more direct feedback from the market or customers, the Product Owner can also invite stakeholders to the Sprint Review. In practice, it is recommended that Developers and Product Owner discuss the completed backlog items before the Sprint Review. In this way, small change requests can still be incorporated in the current Sprint and the effort involved in the Sprint Review remains manageable.

In the Sprint Retrospective, the Product Owner, as part of the Scrum Team, helps to identify potential for improvement in the way the Scrum Team works and to define measures.

Developers

The Developers are responsible for the How. They organize themselves and are responsible for the development of the product. They are responsible for analyzing the requirements, designing, building and testing the product, as well as all the activities that are necessary for the product, such as documentation. As the core of value creation, the developers have technical sovereignty over the product and the product development infrastructure.

In Sprint Planning and Backlog Refinement, the Developers ensure together with the Product Owner that they have understood the Product Backlog items, they estimate the size of the Product Backlog items and provide the Product Owner with feedback on technical dependencies, feasibility, need for revisions, among other things. This is important information for the Product Owner, which in turn has an influence on the prioritization they make in the Product Backlog.

In Sprint Planning, the Developers add tasks for implementing the selected backlog items to the Sprint Backlog as required. They therefore plan the procedure for processing the items.

During product development in the Sprint, the developers work through the Product Backlog items transferred to the Sprint Backlog from top to bottom, as the decision on prioritization still lies with the Product Owner. In the event of problems, this ensures that the least important of the selected PBIs remain at the end of the sprint. Several items can be worked on at the same time, but this increases the risk that many things will not be completed by the end of the sprint. Ideally, all Developers swarm to the first item in the Sprint Backlog and work through item by item. Small items and small teams facilitate this approach and create the agility desired in development.

The Developers meet once a day for the Daily Scrum to plan the next 24 hours. To get started, it has proven useful for everyone to briefly tell the others what they did yesterday, what they are doing today, and where they need support from other Developers.

As soon as an item has been processed, the Developers contact the Product Owner and try to get feedback for this item during the Sprint in order to relieve the Sprint Review. The Developers also contact the Product Owner if there are any uncertainties regarding the backlog items during development or if delays to the planning are foreseeable.

The Developers present the Increment to the Product Owner during the Sprint Review and receive important feedback on the product. They discuss suggestions for improvements to their own working methods with the Scrum Master and Product Owner in the Sprint Retrospective.

Scrum Master

The Scrum Master has no authority within the actual product development, but is nevertheless the key to success in Scrum! Roughly speaking, he has three tasks:

- First, he is an agile coach for the Scrum Team and the organization. This assumes that the Scrum Master has sound experience with Scrum, agility and agile transformations and can help his team, stakeholders and the organization to understand and live agile product development.

- Second, the Scrum Master is the person who de facto ensures the execution of Scrum. He helps the Scrum Team and the organization to adhere to the rules laid down by Scrum.

- And the third function of the Scrum Master is, in my opinion, the most important. An experienced Scrum Team requires very little coaching and Scrum support, which frees up resources for the Scrum Master to focus on his main task: Communicating and removing impediments in the organization that inhibit the Scrum Team from working fast and autonomously.

In recent decades, many product development organizations have lost the sense of how much value creation comes from development and not from the supporting processes. In these organizations, the obstacles that slow down product development are unintentionally created over time. Examples of this can be purchasing and development processes, organizational and responsibility structures and a lack of infrastructure, especially test infrastructure. Figure 19 shows the agile hierarchy: The organization serving the product development process.

"So, the Scrum Master doesn't work, he just looks after his team? Is that supposed to justify a separate position? Can't the Scrum Master oversee several teams so that it is worth it?" I often encounter these questions in Scrum training courses. One possible answer that emphasizes the value of the Scrum Master is, "A good Scrum Master can support two Scrum teams. An excellent Scrum Master, on the other hand, will only support one Scrum Team." What does this mean in everyday Scrum life? The Scrum Team notifies the Scrum Master about the impediments in the organization. These can be brought to the Scrum Master in the Sprint Retrospective, in the Daily Scrum or during the Sprint. The Scrum Master's task is to remove the impediments by communicating the agile principles to everyone involved in the company and gradually changing the organization so that those who create value find ideal working conditions. You may now be wondering whether it makes sense to put product development above everything else. Well, it will when one of your competitors is successful with this concept.

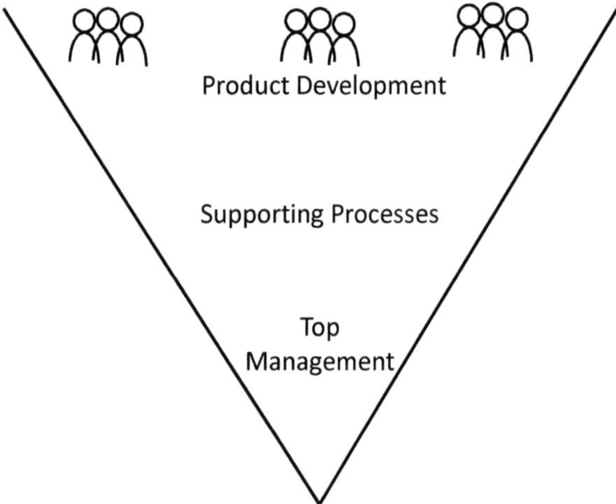

Figure 19: Inverted hierarchy

The elimination of impediments by the Scrum Master is the key to the performance improvements that can be achieved with Scrum.

However, this also requires the willingness of the organization to focus on the Scrum Team as a value creation team, to trust in its autonomous actions and to consciously abandon fundamental elements of traditional project management. The Scrum Master should be well networked in the organization's hierarchy, or the organization should provide an interface in the hierarchy through which he can effectively change things.

As a result, the success of Scrum stands and falls with the organization's willingness to transform, and with the acceptance of the Scrum Master's work. Unfortunately, in practice a Developer is often appointed as a part-time Scrum Master without having the opportunity to effectively help shape or reshape the organization in the interests of their Scrum Team.

Practical tip: Double staffing

The question repeatedly arises as to whether one person can take on several accountabilities in Scrum. In principle, this is possible, as these are merely responsibilities that could theoretically be combined. In my experience, however, there are a few restrictions:

- Developer is also Scrum Master: A setting that I often encounter in practice. The motivation is usually to save money on a freelance Scrum Master. However, this reduces the Scrum Master's activities to the administration of Scrum. As a rule, they cannot then do their main job of eliminating impediments. They usually have neither the time nor the position to effect change in the organization.

- Developer is also Product Owner: This setting is rare. The Product Owner has far-reaching powers with regard to the development schedule and budget – this area of responsibility is rarely taken on by a developer. This constellation also harbors considerable conflicts with regard to the time required for Product Owner responsibility.

- Product Owner is also Scrum Master: I wouldn't opt for this combination under any circumstances. The two responsibilities have fundamentally different interests. While the Product Owner wants as much functionality as possible per Sprint and, in their eagerness, may sometimes question the developers' estimates or exert pressure in other ways, the Scrum Master should prevent precisely that. The distinction creates the opportunity to make conflicts between responsibilities transparent and thus give the Scrum Team the opportunity to learn from them. Transparency, Inspection, Adaptation.

Practical tip: Line functions

The transition to Scrum raises the question of what role the previous team leaders and project managers should take on. Even if the project manager's tasks are now shared across the entire Scrum team, most of the overlap is with the Product Owner. The Product Owner is responsible for content and budget and stays in touch with the stakeholders. The previous project manager must relinquish responsibility for optimizing the working environment and planning and tracking tasks – the Scrum Master and Developers are now responsible for this.

I take a very critical view of using a team leader as a Product Owner or Scrum Master in a Scrum team where the other people are subordinate to them in disciplinary terms. The Scrum Team is designed in such a way that there are no hierarchies or authority within the team – it's a real team.

As a manager, a team leader is in principle a good Scrum Master. They have the position and the network to change things. However, this changes their job significantly, from working in the system to working on the system. Vacation approvals, employee appraisals and other established demotivators do not fit in with the Scrum Master's tasks. For such a setting, I recommend cross-switching the Scrum

Masters between the teams so that a team leader is the Scrum Master of a Scrum Team in which no one reports to them in a disciplinary capacity. Such a setting is difficult if the team leaders have a strong content-related link to their subordinate team. This is relatively common because technical experts are often promoted to management positions due to a lack of other promotion opportunities.

Practical tip: Team WIKI

Many teams use a WIKI or a similar system to record team decisions. Timeboxes, times, tools, procedures, Definition of Done, Definition of Ready and many more can be filed there and is always centrally available to everyone. The Scrum Team decides whether it is to be a WIKI, or whether everything essential is written down on flipcharts in the team room or something in-between.

Scrum Artifacts

Product Backlog

The Product Backlog is the artifact in which all tasks and requirements converge in order to achieve the Product Goal. It is important for me to emphasize the characteristics of a backlog once again. The most elementary: A backlog has a clear sorting according to the priority of the entries, the PBIs. This means that there are never two items with the same priority in a backlog. The item with the highest priority is always at the top, and a backlog is consequently processed from top to bottom. A backlog can include different types of entries. In product development, these are usually requirements, possibly in the form of User Stories (more on this later) or tasks that need to be completed. I will therefore always refer generically to Product Backlog items in the following. Everything that is necessary for the development of the product must therefore be included as an item in the Product Backlog.

Product Backlog items in a backlog can have different levels of detail. PBIs that are at the top of the backlog, and are therefore processed promptly, often have a finer granularity or a deeper level of detail than PBIs that are further down, and are therefore further away on the timeline. Backlog items therefore mature on their way up, so that time and money are not invested too early in specifications and analyses that would be invalidated by later changes.

The Product Owner is responsible for prioritization in the Product Backlog. The entries can also be entered into the Product Backlog by other people after consultation. For example, the Developers can enter a task to revise the architecture or design or a research task by writing these tasks directly into the Product Backlog.

Sprint Backlog

While the Product Backlog is the Product Owner's artifact, the Developers artifact is the Sprint Backlog. In Sprint Planning, they transfer the Product Backlog items to the Sprint Backlog that they believe fit into the Sprint. The selected PBIs strictly retain their priority from the Product Backlog, as only the Product Owner may change the prioritization. Just as the Product Backlog is linked to the Product Goal, the Sprint Backlog is linked to the Sprint Goal, which creates a context for all activities in the Sprint and is intended to bring the project one step closer to the Product Goal.

In contrast to the Product Backlog, the Sprint Backlog does not usually include Product Backlog items with different levels of detail, as all PBIs in the Sprint Backlog must be presented in such a detailed and comprehensible way that the Developers can start working on them immediately and independently.

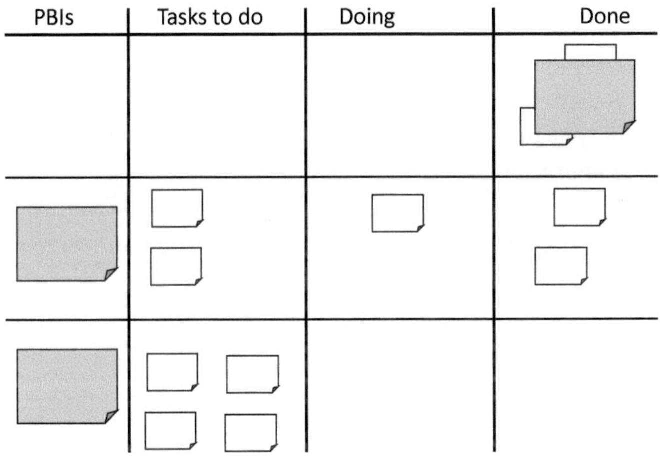

PBIs	Tasks to do	Doing	Done

Figure 20: Sprint Backlog

In addition to inherited PBIs, the Sprint Backlog can also contain tasks that the Developers consider necessary for processing the individual PBIs. These tasks are added to the Sprint Backlog by the developers during Sprint Planning or later during the Sprint.

The Sprint Backlog therefore contains the PBIs planned for the Sprint and the tasks required to process the PBIs (Figure 20). The Sprint Backlog is therefore very often designed as a so-called task board or Scrum board, on which the degree of completion of PBIs and tasks is displayed in various columns.

Please note: Scrum only defines the Sprint Backlog as a backlog for selected PBIs and the associated tasks. The Scrum Board with four columns is a tool that is often used with Scrum, but it is not part of Scrum.

Practical tip: Media selection for Scrum Teams

A single Scrum Team can easily work with sticky notes, whiteboards and flipcharts. This means that the threshold for changing notations and working methods is significantly lower than when using software tools.

Only the maintenance of the Product Backlog would be a little unwieldy with sticky notes, but a small DIY solution with a spreadsheet is often enough in this case.

As soon as several teams work together and/or the team members are allocated, there is no way around a software solution for Product and Sprint Backlogs.

My recommendation: Try to do as much paper-based work as possible at the beginning. This helps to keep the focus on the working method rather than the tool and facilitates quick adjustments.

Increment

In addition to the two backlogs, the third artifact defined by Scrum is of a different nature. It is not a document, but the product that is currently being developed.

The aim of Scrum is to present a new version of the product (incremental development of the product through new functionalities) after each sprint, in a quality that enables the customer to use the product operationally. This is called Done, which is also the shortest existing definition of Scrum: "The aim of Scrum is to deliver a Done Increment." Using Scrum to move the product to a done state every few weeks is naturally easier in software than in system development. However, there are also interesting projects in this area in which, for example, new Increments of a car are delivered every week. The common understanding of Done is documented in the Definition of Done (more on this later).

Scrum addresses two issues with the Definition of Done. On the one hand, the Increment should be able to be used by the customer or stakeholder in the business as soon as possible in order to generate revenue or reduce costs right at the start of the development process. In addition, real use provides the most valuable feedback for the *probe, sense, respond* loops.

However, not every customer wants to go to market with product increments in short cycles, which brings us to the second advantage of Done, which is easily overlooked: Only when the Increment is truly Done and, as a result, all tests have been carried out, all documentation has been updated, and all drawings and source codes have been filed in the relevant systems, can the organization make an objective statement about the progress of the development. I see unfinished business like creases in a carpet – they move, but they stay and grow. It makes it increasingly difficult to predict and therefore increases the unpredictability of development in the organization.

Practical tip: Product Increment

A tested and documented Increment, as in a Done Increment, creates early value for the stakeholder and provides the best feedback on the progress of development activities. In software development, the time required to build and test the product is negligible and the tools for this are established. In system development, on the other hand, it is often not possible to put a tested Increment on the table within a Sprint. Here, the product increment must be defined differently. I use two guiding questions for this:

1. What can we achieve in the next Sprint to add value to the product?
2. What work products can we deliver in the next Sprint, and who can give us feedback on them? (Figure 3)

These two questions are aimed at staying as close as possible to the iterative-incremental concept and not slipping into a waterfall-like approach. A concept or a drawing is worth nothing on its own, only things that have been tested provide feedback. Simulations, experiments, throwaway prototypes, etc. are therefore better than drawings. Ideally, this makes it possible to retain the idea of Increments, but to define Increments at different levels of maturity in order to be able to develop in short cycles.

Scrum Events

Sprint

The Sprint plays a special role among the Scrum events. It is therefore often forgotten in the list of events. The Sprint is a container event that includes all other events (Figure 21). It therefore does not start after Sprint Planning, as is sometimes formulated, but with Sprint Planning and ends with the end of the Sprint Retrospective. As soon as a Sprint is over, the next Sprint starts automatically. There is therefore no period between two Sprints. According to the Scrum Guide, a Sprint lasts a maximum of one month. In practice, the upper limit is often set at four weeks or Sprints are always defined as multiples of a week. This means that the changeover from one Sprint to another can always take place on the same days of the week.

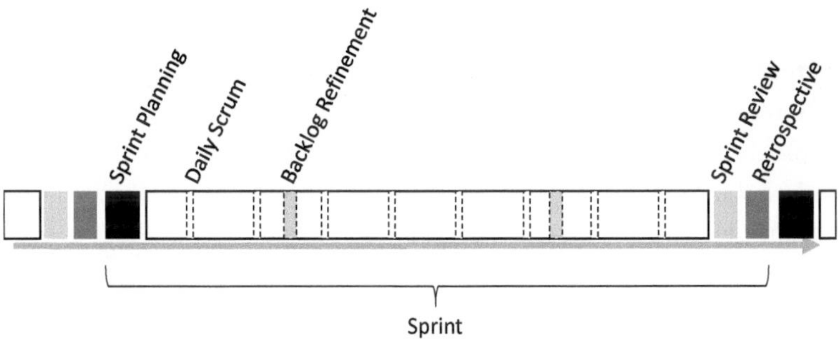

Figure 21: Sprint on the timeline

Practical tip: Moderation

In newly formed Scrum teams, the Scrum Master will take on the role of a moderator. Once the Scrum Team is working well together, the Scrum Master also demands compliance with

the timeboxes and will consistently cancel the events after the timebox has expired.

Sprint Planning

A new Sprint begins with Sprint Planning. The entire Scrum Team
meets, and Product Owners and Developers look at the top entries
in the Product Backlog. When describing Sprint Planning, I assume
that the PBIs are already known and understood by the Developers
and that they appreciate the dimension.

In the first step, the Developers assess what PBIs can be implemented in the current Sprint. In doing so, they must of course adhere
to the prioritization specified by the Product Owner. Consequently,
they transfer PBI after PBI from the Product Backlog to the Sprint
Backlog until the Sprint is full. If necessary, the Product Owner answers open questions about the requirements and, together with the
Developers, defines the overarching goal for the Sprint – the Sprint
Goal.

In the second step – or at the same time as the first step – the Developers plan the activities for the sprint. Tasks that are necessary to
implement the selected PBIs are often written down. These are also
included in the Sprint Backlog. Sprint Planning does not necessarily
have to define all tasks for the Sprint; an unspecified number is sufficient for the Developers to begin their work after this event. The
rest can be defined later, during the Sprint.

The steps described here do not necessarily have to be sequential.
They both have to take place in Sprint Planning – the when and how
is up to the Scrum Team. According to the Scrum Guide, Sprint
Planning for a one-month Sprint should take no longer than eight
hours, and for a standard two-week Sprint no longer than four hours.
In practice, Sprint Planning can be held for well under an hour if all
the Product Backlog items in question have been clarified and estimated in advance.

Daily Scrum

The Daily Scrum is an event for Developers. Scrum Masters, Product Owners or others do not have to attend, but could participate as listeners.

The Daily Scrum is not a reporting meeting, but a coordination and planning meeting for the Developers. The past 24 hours are briefly reflected upon and the work for the next 24 hours is planned. The aim of the event is to provide everyone with the information they need to work independently until the next Daily Scrum. As an introduction to Scrum, it has proven useful for each team member to briefly answer the following three questions:

- What did I do yesterday?
- What am I doing today?
- Where do I need support?

These questions reflect the idea of the Daily Scrum. They were previously included as a suggestion in the Scrum Guide, but have been removed in the current version to reduce the risk of a reporting meeting.

The timebox for the Daily Scrum is defined as 15 minutes, whereby this event does not scale with the Sprint length.

Practical tip: Daily Scrum

In order to master the Daily Scrum in 15 minutes, an intense learning phase is necessary at the beginning, as discussions about solutions must be stopped immediately in order to keep to the timebox. Nevertheless, important discussions must not be ignored. For example, the Scrum Master can interrupt a discussion by handing the main speaker a pack of sticky notes and asking all those involved to make a note of a relevant keyword and stick it on a wall. This wall is then the meeting point for everyone who is interested in the topic and wants to continue the discussion after the Daily Scrum. In practice, more amusing solutions can also be observed, such as holding the meeting in a foreign language to curb the general flow of speech. In any case, a meeting timer (software or hardware) is useful to provide all participants with transparency about the remaining time.

Sprint Review

The Sprint Review takes place as the penultimate event in the Sprint, after the increment has been completed. The review closes the feedback loop for the product. Accordingly, Product Owners and Developers take action here. Stakeholders can optionally take part in the review in order to get better feedback on the needs of the market or internal or external customers.

In the Sprint Review, the Developers demonstrate the Increment to the Product Owner and any stakeholders present. Change requests are added to the Product Backlog. The Sprint is not extended, not even to quickly fix a few minor issues. PBIs that have not been fully processed end up back in the Product Backlog and are prioritized against the existing entries.

The timebox for the Sprint Review must not exceed four hours for a one-month sprint. Depending on the product and type of the demonstration, quite a lot comes together in the Sprint Review. In practice, it is therefore common for Developers to get feedback on individual PBIs from the Product Owner during the ongoing Sprint. On the one hand, this takes the pressure off the Sprint Review and creates capacity for exchange with the stakeholders; on the other hand, early feedback allows small corrections to be made during the sprint. This avoids carpet creases of small change requests that would otherwise be put in the Product Backlog in the Sprint Review. Of course, corrections that would jeopardize the Sprint Goal should be postponed to later Sprints.

Sprint Retrospective

While the Sprint Review, as previously mentioned, closes the feedback cycle regarding the product, the Sprint Retrospective processes the feedback regarding the process. The Sprint Retrospective has a maximum timebox of three hours for a one-month sprint. In contrast to the other events, where silent spectators are welcome, the retrospective takes place behind closed doors offering a safe space for the Scrum Team.

In this event, the Scrum Team reflects on the current Sprint that has just ended. What went well? What didn't? What should be kept? What not? What should be started? The Scrum Team tries to derive potential improvements to suit how they work. After reflecting on and evaluating positive and negative aspects of the last Sprint as systematically as possible, experienced teams agree on a single point for improvement that they can implement in the next Sprint. The Sprint Retrospective, often just called Retro, also offers the opportunity to address and resolve social tension within the team thanks to the safe space.

The following basic rule applies: If people and problems are discussed in the hallway after the Retro, the Retro did not work. Websites such as Retromat (retromat.org) offer various moderation techniques for this event, which is usually moderated by the Scrum Master. When the Sprint Retrospective ends, the current Sprint ends and the new one begins.

Practical tip: Days and times

All timeboxes used, i.e., the Sprint length and the lengths of the other four Scrum Events, are defined by the Scrum Team, as well as the days of the week and times of the events. Most teams schedule the Sprint change – which includes the last two events of the old Sprint and the Sprint Planning of the new Sprint – in the middle of the week. A Sprint end on Friday is difficult for many team members because it takes away flexibility from the working time models. As there is no time between two Sprints, the review and retrospective would have to be scheduled for Friday afternoon in this case. In practice, the events are usually scheduled so that one Sprint ends on Tuesday or Wednesday afternoon and the other begins the following morning. In my experience, a night between two Sprints is better than holding all three events on one day, as it allows you to pause, celebrate and conclude before the new planning phase begins.

The same applies to the slot for the Daily Scrum. In the morning or evening, it would take flexibility out of the working time model, but in the middle of the morning or afternoon, the Daily Scrum pulls developers out of their work. Many therefore schedule this daily event directly before the lunch break, which can also make it easier to meet the 15-minute target.

The exact times for the events are determined by the Scrum Team, and in the case of the Daily Scrum only by the developers. The times should remain constant for a certain period of time, as this significantly reduces the complexity of team planning.

Further aspects of the framework

In addition to the official three accountabilities, three artifacts and five events, the Scrum framework describes other important aspects that I will address in this chapter. In particular, it is about the newly introduced or better situated commitments for the three artifacts, which are intended to define them more clearly: the Product Goal for the Product Backlog, the Sprint Goal for the Sprint Backlog, and the Definition of Done for the Increment.

Product Goal

The connection between the Increment and a vision was included in the Scrum Guide until 2017. In the current version, the vision is no longer addressed, but the Product Goal has been introduced instead. Similar to the Sprint Goal for the Sprint, which has existed for some time, the Product Goal is intended to set a goal and a focus for the entire project and provide the context for the Product Backlog. Accordingly, a Product Goal can be achieved even if not all PBIs have been implemented.

Sprint Goal

In Sprint Planning, Product Owners and Developers formulate the Sprint Goal: A kind of vision for the Sprint. It communicates to the stakeholders, on a more abstract level than PBIs can, what the next milestone is. It motivates the Scrum team by setting a goal that creates a focus for the current Sprint.

Two paths can lead to the Sprint Goal: Either the Product Owner already has a topic for the current Sprint and has moved the corresponding PBIs up in the Product Backlog accordingly, or they define the Sprint Goal together with the Developers based on the current high-priority PBIs.

The better the current PBIs fit together, the higher the focus and productivity for the current sprint, and the easier it is to formulate a Sprint Goal.

If the Sprint Goal becomes invalid – for example, because the requirements have changed radically during the sprint, or technical problems make it impossible to achieve the Sprint Goal – the Product Owner is the only person who has the right to cancel the sprint. In practice, this option is very rarely chosen, as a Sprint cancellation can be very frustrating for everyone involved and could potentially upset the rhythm of the organization.

Definition of Done

The Definition of Done, or DOD for short, defines what must be completed in order to be able to present an increment at the end of the Sprint. The DOD therefore includes the quality criteria for the Increment and is the basis for the transparent measurement of progress. In other words: The DOD provides the context for the Increment.

Aspects of product testing, documentation, and document management are usually defined in the Definition of Done. For example, a minimum DOD for software includes the necessary test types, the requirements for documentation and specifications on the status of the source code in version control. According to the Scrum Guide, the organization specifies the Definition of Done – and if it does not, the DOD is determined by the Scrum Team.

Product Backlog refinement

The Product Backlog refinement is used for the ongoing preparation of the Product Backlog, because after each Sprint Planning the PBIs that have now reached the top in the Product Backlog must be clarified in terms of content, broken down and estimated. As with Sprint Planning, the Product Owner and the Developers are responsible for this. If necessary, other experts or stakeholders are also invited. The Scrum framework does not define backlog refinement as an event, but as an ongoing activity as it does not want to dictate the operational characteristics of the Scrum Team's work on the Product Backlog.

For the Product Backlog refinement, the Product Owner presents the developers with the PBIs that have not yet been communicated and clarifies the requirements in a real conversation. Backlog items often only consist of one sentence. In this case, the Developers ask the Product Owner to find out the specific requirements. Scrum seeks to reverse the obligation to provide. Many things are written down in traditional requirements documents, some of which provide useful information for development, while others are simply a waste of time in the lean sense. The most effective way to define requirements is for the Developers to ask the Product Owner about the exact ideas and check the acceptance of possible solution proposals.

In return, the Product Owner receives information that could influence the prioritization of the backlog. On the one hand, the Developers inform them about technical dependencies between the PBIs, on the other hand, they estimate the size of the PBIs, whereupon the Product Owner can estimate the costs of individual requirements.

Practical tip: Capacity for backlog refinement

In my experience, the effort required to constantly maintain a good quality Product Backlog is often underestimated. My rule of thumb: A Product Owner needs at least half of their working time (full-time position) to work with the Scrum Team and maintain the backlog. In many organizations, this gives the impression that additional work is required. However, the effort involved in clarifying requirements is always there. Scrum only makes this effort transparent and relieves the Developers of this activity to a certain extent.

Calculate the capacity of a Sprint: A Scrum team with five developers has a capacity of ten person-weeks in a two-week Sprint. That's almost a quarter of a person-year, for which you have to prepare the work to be done in such a way that the Developers can get started as autonomously as possible. In scaled environments with multiple teams, the numbers are of course much greater. The largest setting I have ever worked in one calendar day corresponded to two person-years of capacity. This means that several people have to work full-time to maintain the backlog.

Scrum values

In the first book on Scrum, the authors Sutherland, Schwaber and Beedle introduced five values that form the basis for Scrum. Since 2016, these values have also been listed in the Scrum Guide as an integral part of Scrum:

- Focus
- Commitment
- Openness
- Respect
- Courage

In my opinion, when discussing these values, the focus should only be on the interaction of all five values. If individual values are omitted or weighted differently, the whole approach collapses. A memorable example of this is the balance between courage and respect: The courageous person is able to clearly express their opinion to other team members. However, without the balancing value of respect, courage does not lead to better teamwork and therefore prevents better team performance.

Scrum in Use

The Product Backlog

The Product Backlog can be divided into three parts (Figure 22):

1. At the top, the PBIs that are known to the team and are small enough to be processed in a Sprint are listed. These PBIs are referred to as Ready because they could be pulled into a Sprint at any time. The area that is ready usually comprises two to three potential Sprints – this means that finished PBIs are still available to be pulled in if they are processed quickly. Even if the Product Owner is unable to take care of the refinement, there are enough known PBIs available, and the Scrum Team can hold the Sprint Planning without the Product Owner being present.

2. The second part includes PBIs that are known and discussed, but not yet ready. As a rule, these are not yet discussed in the necessary depth and/or are still too large to be included in a sprint. Nevertheless, the dimensions of these PBIs are already roughly estimated so that the Product Owner can make a forecast of when they will be processed. This level of detail is usually carried so far into the future that an estimate of the content and effort for the next release is created – about three to six months ahead – depending on the product and the organization.

3. The third part includes rough ideas for further releases and does not necessarily have to be estimated. The scope of this part has no clear definition.

Please note: To be able to start with Scrum, there only needs to be enough content for the first Sprint in the Product Backlog. The approach in Figure 22 is only a suggestion for dealing with the Product Backlog.

A special kind of PBIs are Spikes: They are research tasks that evaluate the feasibility of certain aspects. For example, a PBI with an uncertain approach to a solution can be preceded by a small amount of basic research and the actual implementation can be postponed to a later Sprint. Such a research task is called a Spike in Scrum, it is estimated normally and the result is also presented in the Sprint Review.

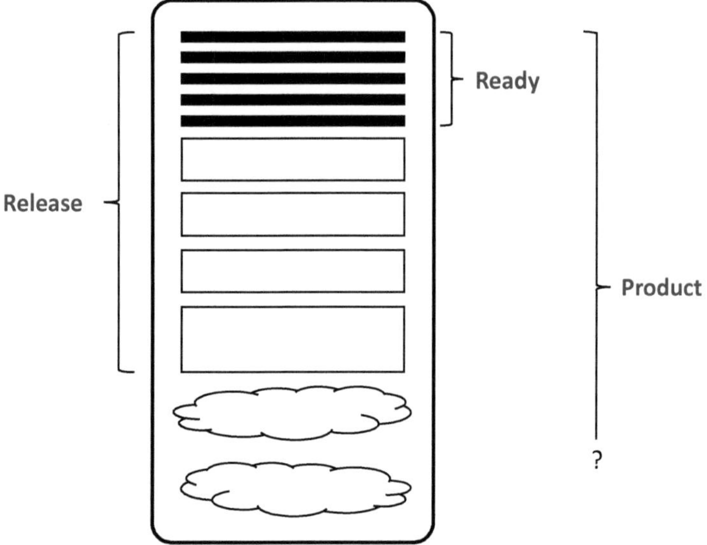

Figure 22: Product Backlog

In practice, there are always discussions about how to deal with the estimates of PBIs if they cannot be completed, and are therefore returned to the Product Backlog. The concept of not changing the estimates of incompletely implemented PBIs has proven itself in the past. As it is not yet clear when this PBI will return to a Sprint when it is moved to the Product Backlog, it makes sense to retain the estimate for the time being. If you don't touch the PBI for another six months, you are mentally starting from scratch, which means that the original estimate will probably fit. If, on the other hand, you use the PBI immediately in the next sprint, the Developers can of course adjust the estimate downwards in Sprint Planning or Backlog Refinement to take into account the work already completed.

Practical tip: Differentiation between PBIs and tasks

A few rules of thumb for a good split between Product Backlog items and tasks:

Product Backlog items
- include the What,
- are often requirements,
- are largely independent of each other and can be swapped in their order, and
- are addressed to the entire Scrum Team.

Tasks
- include the How,
- are technical or administrative tasks to implement the corresponding PBIs,
- are often interdependent and must be processed in a specific sequence, and
- often address individual developers or subsets of the Scrum Team.

Definition of Ready

When is a Product Backlog item ready? The Scrum Team determines this in the so-called Definition of Ready (DOR). The DOR is the working basis for the backlog refinement, and for the Product Owner when they change the priorities in the Product Backlog. As previously described, the top items in the Product Backlog should always be Ready, and correspond to the DOR.

In my opinion, a minimal DOR includes three criteria:
- The backlog item has been understood by everyone
- Its dimension is estimated
- It is not greater than xy (so that several items fit into one sprint)

The Scrum Team can add further criteria depending on requirements, product and organization. It is important to design the DOR precisely, so that the developers can work on the items. However, the quality criteria for the PBIs must not be set so high that the agile approach turns back into phase-driven development with a long analysis phase. The DOR can be adapted by the Scrum Team in the Sprint Retrospective based on the experience gained. Ideally, it includes just a few key words or short sentences and is documented in a way that is visible and accessible to everyone involved. In the simplest version, the DOR is presented on a flipchart in the team room or is filed on a server in the shared team WIKI.

Relative estimates

The agile approach of no longer estimating tasks or requirements in units of time is one of the most discussed points in agile training because it strongly contradicts our socialized opinions and methods. Historically, however, the move away from person-hours and person-days does not come from the agile world, but is based on a research contract from the US Department of Defense. In the 1940s, it commissioned the Rand Corporation think tank to find out how projects could best be estimated. The three key messages were
- Estimates in units of time are very error-prone
- People can estimate relative values more easily than absolute values
- Experts should estimate independently in order to avoid the anchor effect known in psychology

However, in what unit can activities be estimated in if not in hours or days? Jeff Sutherland uses a journey as an analogy. Let's say you want to get from Frankfurt to Berlin. How long will it take you? This is not easy to answer, as there are many factors that play into the answer. The question of the distance between Frankfurt and Berlin, on the other hand, is easy to answer: It is about 400 km as the crow flies. The time required depends on the means of transportation and the route, and if you know your speed, it is easy to estimate the time well. If it is estimated in kilometers, it does not need to be corrected if you are traveling at a different speed from then on. Similarly, progress measurements – How much further is it? – are more objective in kilometers than in hours. This is because stating the remaining time in hours is only reliable if you can maintain your speed.

The transfer to the Product Backlog and the estimation of tasks in general raises the question: What is the absolute size of a task? What analogy from the travel example applies? There is no unit or measurement system for this. This is where the second aspect from the above-mentioned study comes to the rescue: People are better at estimating relative sizes than absolute ones. This means that we can estimate the size of tasks relative to each other, without dimension. For conversion into units of time we need, as in the travel example, a second input variable: Current velocity.

In agile estimates, Estimation Points are used as a unit for the size of tasks. A reference task is assigned a value and all other tasks are estimated in relation to it. Is a task roughly twice as big, four times as big or half as big as the reference task? This approach is initially unfamiliar in practice, but with a little practice it is much quicker than estimating in time units.

To be able to make statements on the time axis, you need the velocity of the Scrum Team, measured in estimated points per Sprint. At a first glance, working with size and speed may seem somewhat complicated compared to estimates in time formats. However, even with time estimates, the available capacity must be determined in order to define the scope of a Sprint. This is just as error-prone as a speed forecast for a car. The background noise of meetings, training sessions, vacations and so on, you must determine how long a task estimated at eight hours, for example, will actually take on the timeline.

In practice, the effort required to determine this capacity is disproportionate to the achievable accuracy. A speed based on estimated points, on the other hand, does not have to be elaborately calculated. It is determined empirically and can therefore be estimated much faster and more precisely.

In addition to a faster estimation process and more accurate estimates, estimates of size have the significant advantage of being very stable, as they are independent of the achievable speed. This means that they are independent of the people working on the tasks, independent of the quality criteria that are applied to all tasks, and independent of experience and learning curves of those involved. Estimates in time units, on the other hand, would have to be adjusted every time the framework conditions changed: More experience in the technology — all estimates in the backlog would be adjusted downwards. More test coverage for all PBIs — all estimates in the backlog would be adjusted upwards.

To conclude the consideration of absolute and relative estimates, I would like to add a psychological aspect: Those who estimate in hours always see themselves monitored in the execution of the tasks, whether by others or by themselves. If the actual execution deviates significantly from the estimate, this always provides an opportunity for personal criticism. This is an effect that does not occur with estimates using relative values, as the estimate is deliberately decoupled from the time axis.

Practical tip: An anecdote about estimating

Doing away with time units when estimating is not easy and is difficult for many people to process. Perhaps this short story will help you understand more:

One Scrum Team I coached decided to estimate in hours, not points, when they started Scrum. That's fine, it's the team's decision. So, in the first Sprint Planning phase, the Scrum Master went to the whiteboard to determine the capacity of the developers. All the Developers went through their calendars, and each of them confirmed the number of hours they could contribute in the next Sprint. The Scrum Master noted everything down and did the math: 168 hours. So, the Developers pulled Product Backlog items from the Product Backlog into the Sprint Backlog for just under 168 hours. The work began. In the Sprint Review, the realization came to light: about two thirds of the Sprint Backlog had been completed, the equivalent of 110 estimated hours. Not an issue as this is normal for the very first sprint.

The procedure began again in the next Sprint Planning. The calculated capacity this time: 174 hours. The result of the review: 108 estimated hours completed. In the third Sprint Planning, I intervened in the discussion. Before the Scrum Master could write down his list, I called out 110 hours to the group. The team was horrified. They had calculated 170 hours. A considerable difference between the exact calculation and my gut estimate. The Sprint then ended with 112 completed estimated hours. The team was obviously able to deliver around 110 of the estimated hours quite reliably. The fact that this does not match the calculated capacity may be due to estimation errors on the Product Backlog items or estimation errors in the capacity. To find this out, all the working hours would have to be logged precisely.

Do you recognize the system? My estimate of 110 hours was very accurate every time. It required no effort because it was based on empirical data from the last Sprints. However, it actually had nothing to do with hours. The PBIs were numbers of estimates and we could easily determine what was a realistic amount of work for the team. Actually, it had long been an estimate in points, with velocity measurement. Sticking to hours and the team's capacity calculation was a lot of work and significantly reduced the quality of the estimate.

Estimation scales

Whether you estimate relatively or absolutely, there is another interesting aspect regarding estimation accuracy for different sized Product Backlog items: People can estimate smaller items more accurately than large ones. Remember the travel example from the previous section? You can estimate the duration of a short walk from you to your neighbor relatively easily. It is probably a little further from your home to the nearest bank and the estimate will be correspondingly less accurate. At the latest, when we get to the next question, "How long does it take to walk from Frankfurt to Berlin?" you will ask me with counter-questions, as the requirement is not clear. What about breaks and overnight stays? Even if the requirements are clarified, the estimate for this very long walk is likely to be quite inaccurate.

To take this fact into account, scales that spread upwards are used for estimation. The Fibonacci series, which corresponds to natural cell growth, is widely used. The Fibonacci series is created by starting with the numbers 1 and 1 and always forming the next number from the sum of the two previous numbers. The series is therefore 1, 1, 2, 3, 5, 8, 13, 21, 34, 55, 89...

If this series is used for estimates, it is not possible, for example, to estimate a PBI to the value 75, as this would simulate a non-existent accuracy. However, the neighboring values 55 and 89 may be used and are just as correct or incorrect as 75. Many teams use the Fibonacci series by defining the values from 1 to 8 as ready and using the values 13 to 89 as rough estimates for later refinement. This has the advantage that two series of numbers are created whose columns are connected by a factor of approximately 10 (Figure 23 above).

In addition to the Fibonacci series, a Fibonacci series modified by Mike Cohn is also widely used. Cohn has replaced the values above 13 with the round numbers 20, 40 and 100. In this series, estimates that are large and imprecise can be recognized by the zero in the number (Figure 23 below).

| 1 | 2 | 3 | 5 | 8 |
| 13 | 21 | 34 | 55 | 89 |

| 1 | 2 | 3 | 5 | 8 | 13 |
| 20 | | 40 | | 100 | |

Figure 23: Estimation scales

Estimation methods

The Rand Corporation study stated that relative estimates require that the experts involved to make their estimates independently of one another. Furthermore, it makes sense to include all participants in the estimation process, as in practice it is often observed how experts make estimates and these are accepted without reflection by other participants by virtue of their expert status. Unfortunately, this means that many opinions remain unconsidered, not only with regard to the estimated figure, but also with regard to the clarity of the requirements.

Even if all Developers in a Scrum Team participate in the estimation, the problem of the anchor effect mentioned in the study remains. This occurs when a person is subconsciously influenced by numbers they have recently read or heard when estimating something unknown and positions their own estimate close to the anchor number. Consequently, the optimum is a secret estimate by all Developers. The best-known estimation method that fulfills this requirement is Planning Poker.

Planning Poker

Planning Poker cards consist of one of the above-mentioned series of numbers. Most card sets available on the market use the Fibonacci series modified by Mike Cohn (Cohn, 2005). In such an estimation, the Product Owner presents a PBI. Each Developer quietly considers an estimate for themselves and selects the corresponding poker card. All participants place the selected cards face down in front of them to avoid anchor effects. The cards are then turned face up, the largest and smallest estimates are sought, and the corresponding participants explain their estimates. The differences here usually arise from a different understanding of the requirements or solution ideas. While the participants then discuss the estimation results, the Product Owner is available to answer questions about the requirements. This is followed by a new estimation round. If large deviations occur again in this round, the Developers must agree on an estimated value. Neighboring numbers in the number series are not a problem, as they are to be considered identical due to the inherent imprecision of an estimate.

Holding more than two rounds of poker for a PBI takes a lot of time, but no participant should be outvoted by the majority. It therefore takes some practice to quickly get an estimated value from the discussions that all developers can support. In my experience, the biggest hurdle is to give up insisting on one's own point of view in order to achieve an estimation result that makes sense for the team in terms of accuracy and estimation speed.

For me, Planning Poker is first and foremost a tool for detecting unclear or misunderstood requirements, which delivers an estimated value as a by-product. The quality of the results in Planning Poker is very high, but the process is complex, which is why it is often only used in Product Backlog Refinement for individual PBIs.

Magic Estimation

If a Product Backlog is to be estimated initially, other methods must be used because Planning Poker is too time-consuming for this purpose. One method for quickly and roughly estimating many PBIs with many people is the Magic Estimation, which is used in slightly different ways. I present one possible approach below.

The PBIs to be estimated and known to the Developers are distributed evenly to all team members as index cards or printouts. An axis is defined on a long table or on the floor, on which the items are positioned according to their size. Now the participants begin to place their cards in turn, whereby cards that have already been laid can be moved to accommodate your card in between. However, the order of the cards already in place may not be changed. Once all the cards have been placed, the second round begins. Each participant can move cards that they think are in the wrong position. The Scrum Master, Product Owner or the involved participant use a pen to mark the cards that are moved by placing a dot or dash on the card for each move.

After the second round, the cards with the most markings, i.e., the most movements, are discussed by the Developers and repositioned if necessary. In contrast to Planning Poker, the discussion is therefore limited to a few conspicuous PBIs, the majority remain in their position without further discussion. Magic Estimation is therefore significantly faster than Planning Poker, but can never match its quality in terms of clarifying requirements and estimation accuracy.

Up to this point, the cards are merely sorted by size, without a concrete assignment of estimated values. In the final step, the Developers position a scale with the number series used for the estimate next to the cards. Planning Poker cards are often used for this purpose. When creating the scale, several PBIs can be grouped together if they are to receive the same estimated value.

You can find tips and tricks on estimation methods on the Internet, where the agile community constantly publishes new ideas and findings.

Progress measurement with burndown charts

The use of burndown charts can be traced back to Jeff Sutherland's career as a pilot. Legend has it that his colleague at the time wanted a precision landing for his project instead of – as with all previous projects – overshooting the target date.

Jeff, the former F-4 pilot, knew that a precision landing is decided by a precise approach. A few knots too fast or a few feet too high and the jet overshoots the runway and crashes in the forest or in the desert.

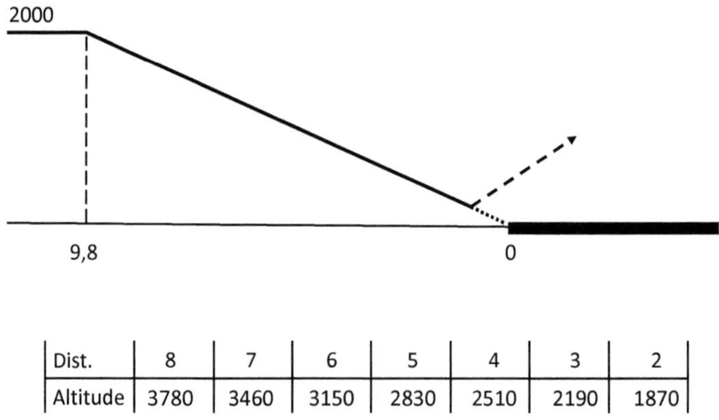

Dist.	8	7	6	5	4	3	2
Altitude	3780	3460	3150	2830	2510	2190	1870

Figure 24: Approach profile

As a pilot, I would like to take a closer look at the history of the burndown chart. An exact approach requires an exact approach profile: A target altitude is defined at intervals of one nautical mile. The pilot flies this defined approach profile by checking his altitude against the target value mile by mile and making small corrections (Figure 24). Sutherland applied this concept to projects, using the remaining worklist as the flight altitude and defining the checkpoints on a time axis.

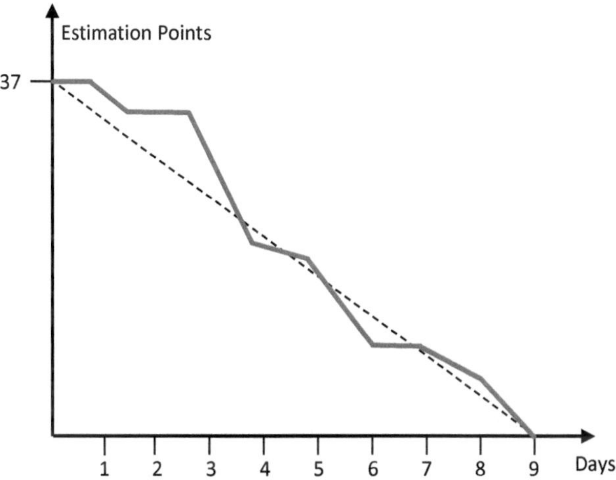

Figure 25: Sprint burndown chart

Sprint burndown chart

The most important burndown chart is the Sprint burndown chart. It covers the duration of a Sprint and the individual days are plotted below as checkpoints. The initial flight level is the sum of the estimates of the PBIs in the Sprint Backlog, i.e., the amount of work that needs to be completed (Figure 25). The ideal line corresponds to a linear decrease in the remaining work over the Sprint. The remaining work is plotted every day, creating a line over the course of the Sprint. Only fully completed Product Backlog items are taken into account.

The Sprint burndown chart is a tool for Developers. It is not an external reporting tool. It is used by Developers to take early counter-measures in the event of problems or, if necessary, to warn the Product Owner if it is foreseeable that not all chosen PBIs can be implemented in the Sprint.

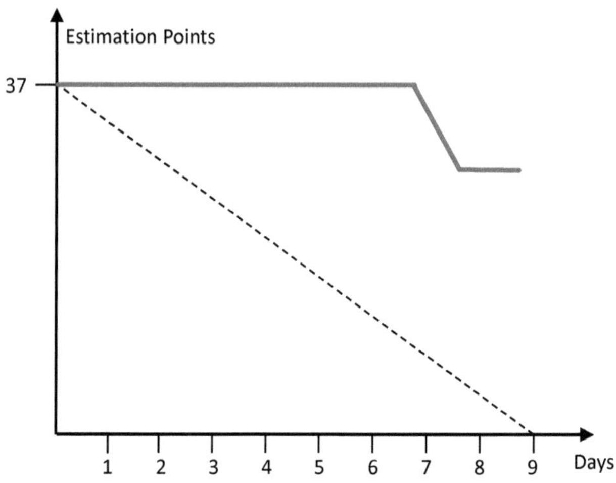

Figure 26: Sprint burndown with flatline

Typical patterns

There are various typical patterns in this diagram that provide information about challenges and potential in the Scrum team. A straight line without a decrease in the worklist (Figure 26), known as a flatline, can have various reasons: Either the Scrum Team only has one big PBI in the Sprint and there is a sudden burndown on the last day – or not. Or the team has technical and/or personnel problems and cannot continue working on the current PBI. If the Developers have worked on too many PBIs at the same time and none of them have been completed, this can also be the reason for a flatline. A flatline is therefore not a one-dimensional diagnosis, as it can have various reasons.

Flatlines are not uncommon in new Scrum Teams, so the burndown diagram offers the Scrum Team an easy way to optimize their work. In particular, it motivates them to make the PBIs smaller. In this way, more items in a Sprint and a daily change in the diagram becomes visible. This is the only way to ensure a precise approach to the end of the Sprint.

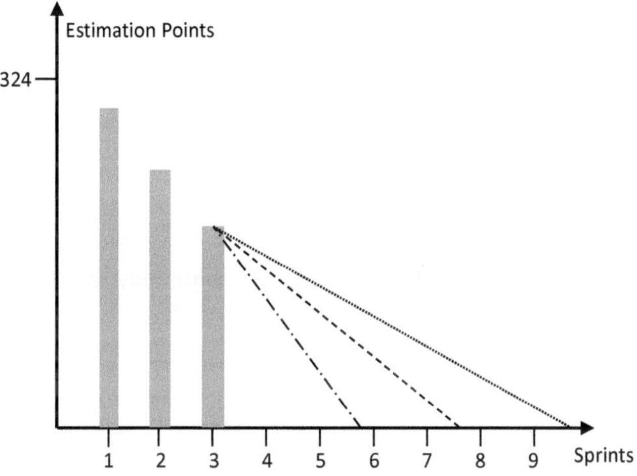

Figure 27: Release burndown chart

Release burndown chart

A second important burndown diagram is the release burndown chart. The Product Owner can use it to measure progress over several Sprints, usually up to the next planned release of the product. Sprints are plotted on the horizontal axis as checkpoints for the flight level. The vertical axis shows the sum of the remaining estimates for the next release (Figure 27).

The image shows that if the Product Backlog remains unchanged, the difference between two adjacent bars corresponds to the speed of the Scrum Team. After at least three Sprints, the Product Owner can estimate when the PBIs planned for the Sprint will be completed, for example, by entering the an extrapolation with the best, worst and average team speed in the last Sprint bar. This gives the Product Owner an estimate of where the completion date for the release will be.

This takes into account the fact that product development cannot be planned for a specific completion date without sacrificing quality or adding time buffers – both very expensive options to meet a given deadline. In the Technical Debt chapter, I will explain why quality is non-negotiable in Scrum.

In principle, the Product Owner therefore has two options for his release planning: Either communicate a specific scope for the up-coming release and leave the date open, or promise a specific release date, but leave the exact content open. If, on the other hand, the Product Owner has to commit to the scope of a project by a certain date, a buffer in the form of additional sprints must be planned in. This approach is avoided by Scrum – if possible – because buffers are expensive. They are simply an exchange of money for probability. Do you agree that this is a good conclusion to this section?

Practical tip: Milestones and agile development

Time and again, I experience uncertainty when it comes to the question of how milestones relate to agile planning and measurement.

Some even think that these two elements cancel each other out. My opinion: The opposite is the case; agile product development requires a milestone plan. Using the method described above with release burndown charts, you can work towards your milestones in a targeted manner. The milestones should not be more than three to six months apart.

The key difference in the agile environment: Milestones should always describe product states or increments, not abstract paper states such as finished concepts, approved requirements and so on. In the agile world, a tested Increment is the only valid criterion for assessing progress.

User Stories as Backlog items
User Stories are a widely used tool in software development for dealing with requirements. They support the lean concept of investing time and therefore money in defining requirements as late as possible in order to avoid waste. A user story initially consists of just one sentence, which could be structured according to the following pattern

As <role>
I would like <goal/desire>
to <benefit>

Example: As a service technician, I would like to receive a report on the machine running times in order to be able to assess the current wear.

In contrast to classic requirements, user stories include two essential additional elements: The actor and the motivation for the requirement. The actor is described either as a defined role or as a persona. This allows Developers to assess what knowledge and skills the actor has and better tailor the implementation to them. You can find out more about personas, for example, at en.wikipedia.org/wiki/Persona_Human-Computer-Interaction.

The motivation for a requirement is important in order to distract from the often-encountered formulation of a solution and to give the developers the opportunity to propose alternative solutions to achieve the goal. This often makes it easier to fulfill the actor's actual wish than the formulated requirement would suggest.

The lightweight approach mentioned at the beginning is illustrated by the life cycle of a User Story. Such a Product Backlog item goes through three phases, which you can remember with the help of the CCC acronym:

- Card
- Conversation
- Confirmation

At the beginning, the User Story only consists of a card with a sentence, according to the pattern mentioned. This story does not incur any significant costs, but is still a very rough requirement. In the agile world, this card is referred to as a promise for a conversation in the future.

If the User Story continues to move up the Product Backlog, it will be discussed in the backlog refinement. This corresponds to the second point Conversation: The Developers identify the requirements conversation with the Product Owner. These are written on the back of the respective card in the form of acceptance criteria. Acceptance criteria document the Product Owner's expectations for this item in a User Story or what they want to see in the Sprint Review in order to be able to confirm the desired implementation of this item. This is the last level: Confirmation.

User Stories are generally only used when people interact with systems. However, it can sometimes be observed how technical requirements are also pressed into the format of User Stories under slight duress: As a microcontroller, I want a supply voltage with a ripple of max. 2mV so as not to get out of step. Personally, I wouldn't put this as a User Story, but as a classic requirement in the Product Backlog.

Quality is not negotiable

Quality must be created in the process; this is a core aspect of the Toyota Production System. Adding quality later is always the more cost-intensive way. Applied to Scrum, this means that every increment must always be of perfect quality, as there must be no carpet creases from pending tests and outstanding documentation. According to Scrum, every Increment must be potentially shippable to the customer. There are two reasons for this requirement:

- On the one hand, the customer should be able to use the product with its core functionality in their business at an early stage. This enables them to generate sales or reduce costs, while real-life use provides the best feedback for further requirements.
- On the other hand, only an increment without postponed activities can provide reliable feedback on the progress of the project.

I would like to present the following model that relates to the key topic of quality: Product development must try to keep the three parameters of quality, functional scope, and delivery time together (Figure 28).

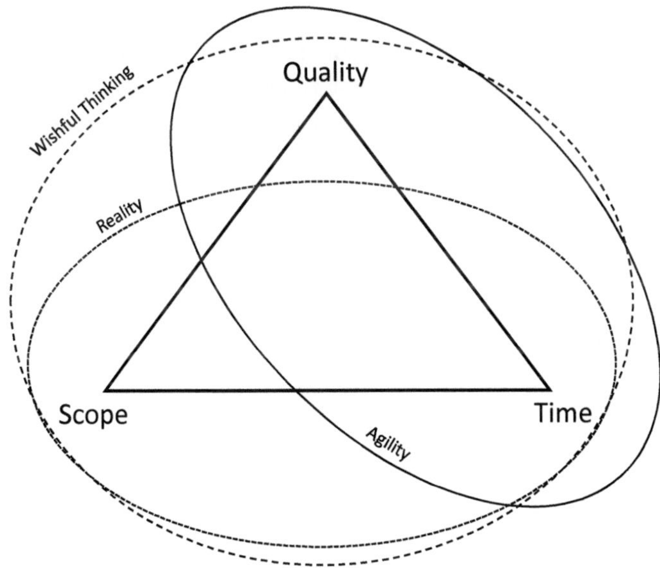

Figure 28: Quality vs. scope and time

The idea of adhering to all three parameters without compromise is pure wishful thinking, as is clear from my previous comments. Product development includes a large degree of variability to which we have to react. In the past, scope and time were often fixed and non-negotiable. This means that the development team only has the option of making allowances for imponderables. I am not referring to the external quality that a customer may notice, but to the internal quality. In practice, variations are made here, for example, by working past the architecture or saving on documentation or postponing it until later.

These postponed activities make subsequent maintenance of the product more difficult and more expensive. Whether they are made up for or not, postponed tasks will be paid for in the future, at a multiple of the current costs. Or, as it is formulated in the agile world: The organization takes on technical debt that must be repaid in the future with compound interest.

In the agile world, quality is non-negotiable. This is the only way to minimize costs over the product life cycle. Therefore, in order to meet the reality of product development, the parameters *time* or *scope* must be varied, as already explained in the release planning. On the smaller planning cadence, the Sprint, Scrum sets the time and leaves the scope open as a variable.

Once again, the variability that inevitably exists in product development, does not make it possible to adhere to the three parameters of *time, scope, and quality* at the same time. Cutting back on quality inevitably leads to an increase in costs. Scrum uses the *scope* parameter as a variable. The time remains fixed in order to achieve maximum learning effects when planning and estimating in identical framework conditions.

Technical debt

As mentioned above, postponed architecture optimizations or redesigns, postponed documentation and postponed tests should be regarded as debt. The debt must be repaid with compound interest – either by making the product more difficult to maintain orby later catching up on the product, which will have become more complex by then.

Avoiding technical debt and reducing it is the accountability of the Product Owner. The Developers generate the necessary tasks and specifications, re-designs or refactorings, and place them in the Product Backlog. However, the Product Owner ultimately decides when these are to be processed by prioritizing them in the backlog.

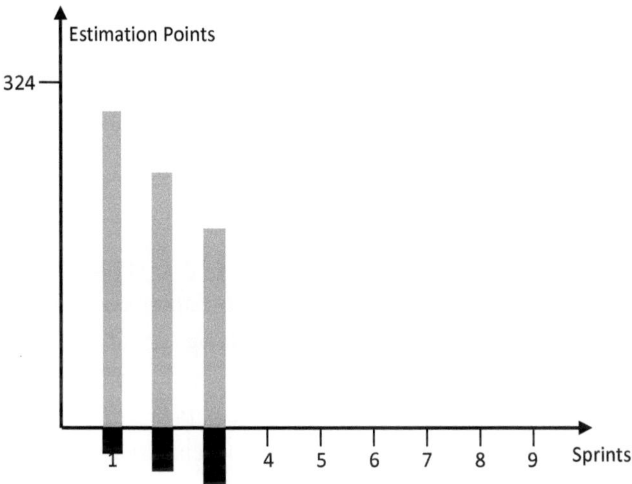

Figure 29: Technical debt in a release burndown chart

Technical debt causes interest, because the later refactoring is tackled, the more expensive it becomes. If it is not tackled, it becomes increasingly difficult to maintain the product. Particularly, in software development, it is often the case that a product is entirely re-developed after many years because the current status is no longer maintainable.

Technical debts are therefore tasks that need to be completed in addition to the functional requirements. They can be drawn below the axis in a Sprint burndown chart (Figure 29). This is an accurate depiction of reality, as technical debts are invisible, so below the surface, but remain part of the work to be done and increase continuously, even without the product being developed further. As the Product Owner is commercially responsible for the entire product life cycle, they are well advised to constantly keep an eye on the identified technical debts and prioritize their processing accordingly.

However, reducing technical debt does not always have to be the top priority. If necessary, the Product Owner can deliberately postpone an internal redesign, i.e., take on technical debt, if they need a functional boost for a customer meeting or a trade fair. However, they should be aware that a redesign carried out later will be more expensive than one planned immediately.

Practical tip: Certification exams

Two different organizations offer certification exams for all Scrum responsibilities: the Scrum Alliance and scrum.org. The exams take place online and are available for different levels of expertise.

Status 2024: At scrum.org, exam participation can be purchased without necessarily having to attend an official training course. At the Scrum Alliance, on the other hand, taking an exam is linked to participation in a corresponding training course. The scrum.org certifications are valid for life, whereas the Scrum Alliance certificates have to be renewed every two years. In my opinion, there are no relevant differences in the market value of the certifications. Many participants decide according to the available training courses. With the knowledge in this book, you can definitely approach the Scrum Master certification well prepared. Good luck!

I do not care what you call it ... whatever
the technique is called, if it's a good one,
it comes down to two things: a commitment to respect people and a commitment to constantly improve.
Matthew May

Kanban in Development

Kanban boards

History

The implementation of Kanban in development was significantly influenced by David Anderson. As a project manager at Microsoft in 2004 and 2005, Anderson began using pull systems from production for software development. He and his team were involved in maintenance projects and had the problem that his system was clogged — not with material, but with tasks. Anderson was familiar with Goldratt's Theory of Constraints, and Drum-Buffer-Rope control. As mentioned earlier, TOC is more widespread in North America than in Europe. Anderson's approach, like the flow of materials in manufacturing, was to create and control a flow of maintenance tasks in his Microsoft team.

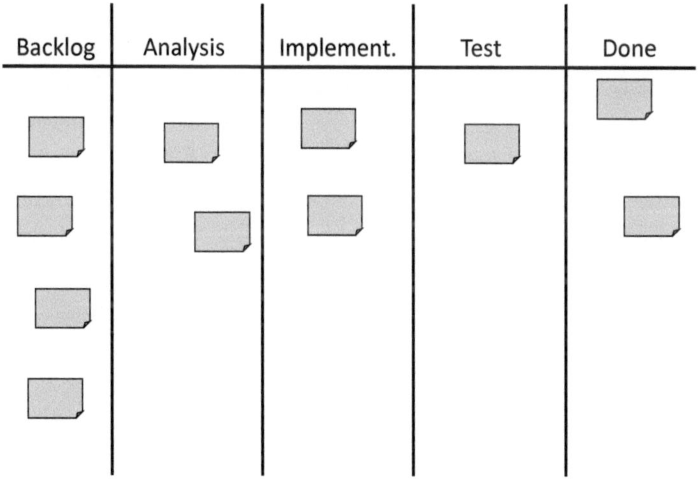

Figure 30: Kanban board (schematic diagram)

The first step in doing this: Visualize tasks with sticky notes on a whiteboard (Figure 30). Anderson had mapped the process for the tasks in different columns on the Kanban board. The cards flow from left to right through the individual process steps.

Anderson quickly achieved initial success with his experiments in terms of throughput time, as the use of DBR reduced the number of simultaneously processed tasks (WIP). However, DBR was difficult to handle in the software maintenance project. Unlike in production, the drum, the planning of the bottleneck, was not practicable to implement in this volatile environment.

At a conference in 2006, Anderson presented his experiments with the whiteboard to the public and was told by Donald Reinertsen – you see, it's a small world – that Kanban control would be better suited to his purposes. The result was Kanban boards as we know them today, with pull control from process step to process step. David Anderson was able to reduce throughput time by 90 per cent within nine months using this approach (Anderson, 2011).

Structure

A Kanban board according to David Anderson depicts the process or an abstraction of the existing process steps in the columns (Figure 31). The tasks, noted on cards, are moved from column to column according to the Kanban system. Unlike in production, there are no transport containers on the board that limit the number of simultaneous tasks (WIP). A separate WIP limit is therefore defined for each process step: There may only be as many cards in the process step as the limit noted in its header. A Kanban stock is defined for the transfer from one step to the next, by dividing each column into two sub-columns. Cards in the first sub-column (Doing) are currently being processed. Once the respective process step for the card has been completed, it is moved to the second sub-column (Done). This is the transfer point for the subsequent process step: From there, the card is removed as soon as it has capacity, i.e., it is below the WIP limit. The illustration shows that the last step has no such sub-columns, as it can pass a card directly to the Done column of the board after processing. Different steps can have different WIP limits. It is important to know that the limits always apply to the sum of both sub-columns. Completed tasks also count towards the WIP of a column.

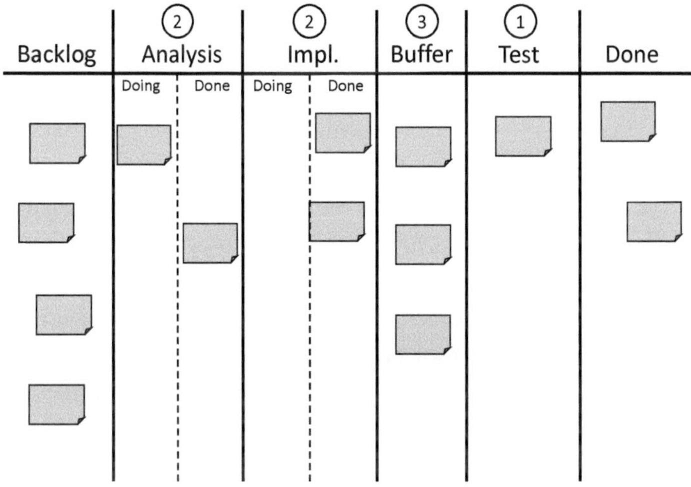

Figure 31: Kanban board with buffer

When switching from DBR to Kanban, David Anderson retained DBR's concept of bottleneck buffering. So, there is a buffer column on his board in front of the assumed bottleneck. The buffer column is a parking space with its own WIP limit to ensure that the bottleneck is constantly supplied with tasks. However, this concept from production only works in situations where the bottleneck is reasonably stable in one place. This was the case with Anderson in maintenance. In development, however, the situation of changing bottlenecks often arises: Under certain circumstances, each task that is dragged across the board experiences the bottleneck at a different point in the process.

Use of Kanban boards

Here is an example of how to work with Kanban boards: All tasks to be completed are placed as a card in the first column, the backlog. This column does not yet have a WIP limit; this is where you collect what needs to be Done. If required, this column can be preceded by other statuses in order to show the maturity of the tasks that have been entered.

The next column, the ToDo column (also called Selected Backlog or Input Queue depending on the source of the literature) is used to define the next important tasks by pulling them from the backlog into ToDo. This column already has a WIP limit; you can see this by the number in the column header. In our example, a maximum of four cards are allowed here. Someone in the company must therefore be found to decide which four tasks are to be tackled next. I deliberately phrased it this way because it is often not that easy. This decision-maker or a committee now places the four cards that are most important to them in the ToDo column, or pulls a card if a space has been freed up in ToDo.

Developers working in the Dev column can now pull a maximum of two cards in the Dev/Doing column. The cards, that they have completed are moved to Dev/Done, i.e., to the handover point. Assuming that the developers have completed a task, the board then looks like Figure 32.

As described, the WIP limit of the Dev column applies to both sub-columns. In the previous example, developers are therefore not allowed to pull a third card in Dev, even though they have already processed one card. This may sound confusing at first, but the core of all pull systems with a WIP limit is that problems can lead to the

                    Kanban in Development

entire process being stopped. Remember the worker at Toyota who is allowed to stop the whole line if there is a problem?

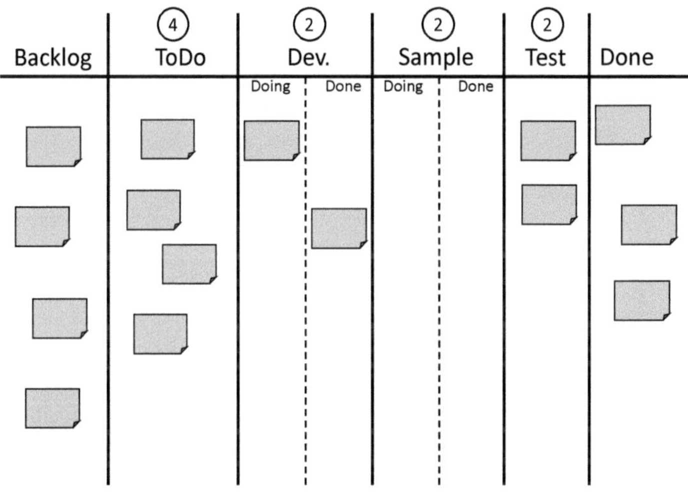

Figure 32: Example for a Kanban board in mechanical engineering

In our example, it can only continue if the team in the Sample column pulls the card from Dev/Done. The sample construction team can do this immediately, as it has not yet pulled a card, and does not violate its WIP limit due to the pull from the upstream column.

If problems occur at one point in the process, a process step cannot take any more cards from its predecessor due to the WIP limit. If you take this methodology further, you can see that problems propagate in the form of blockages from the location of the problem on the board to the left, i.e., upstream. Many people who hear about this concept for the first time are unsettled. This is because it is obvious that such blockages can occur and result in some employees being unable to continue working. This provides the necessary transparency to constantly optimize the organization in small steps, because

WIP limits prevent problems from being circumvented instead of solved. In terms of lead time: If you remember what I said about Cost of Delay, you might know that the costs of blockages are generally lower than the costs of high WIP and the resulting queues.

More Kanban

Type of flow

In the previous chapters, I described how Kanban boards map the process. This is useful for finding the bottleneck, planning buffers, and optimizing the process — as implemented by David Anderson. I will refer to such a Kanban board as a process board below. However, not all processes are as linear as Anderson's maintenance processes. Process boards assume that the process steps are highly sequential and that the majority of activities are mapped in this process (Figure 33). Individual requirements, bug fixes or change requests pass through the stages on such a board until they are implemented or processed.

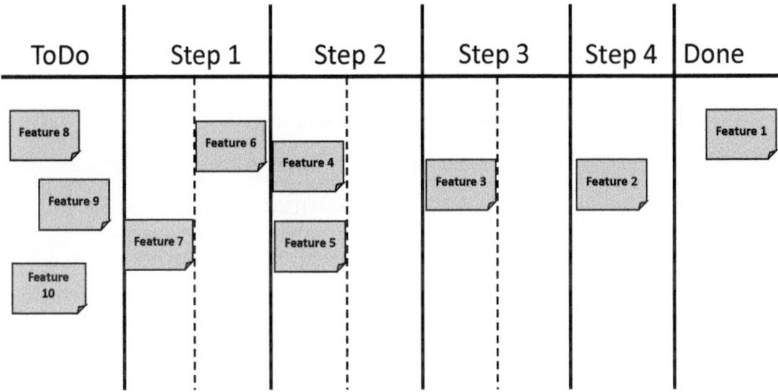

Figure 33: Process board

If you want to set up a process board, you must first decide what is represented by the individual cards, i.e., what flows from left to right. Since the activities are in the columns of a process board, aspects of the product must flow across the board as cards, i.e. requirements, features, assemblies, components – whatever always goes through the process in your setting.

In electronics and mechanics development, the requirements are often interdependent. The integration stages in these domains usually only make sense if a certain number of flowing elements have come together. A one-piece flow of requirements on a Kanban board is therefore difficult to implement in mechanical and electronics development.

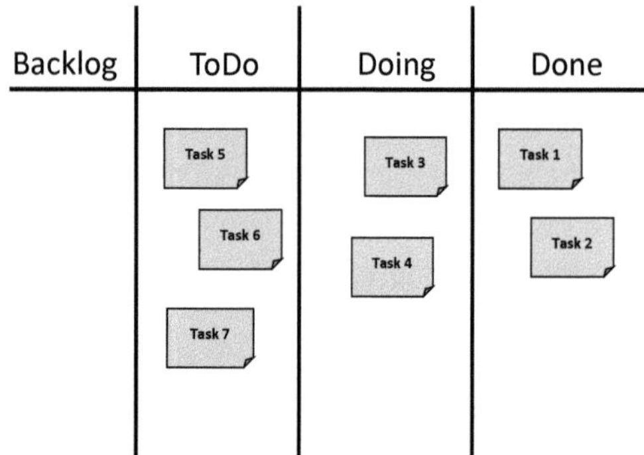

Figure 34: Task board

However, in order to benefit from Kanban mechanisms in practice, a task board, similar to a scrum board, is often used for complex processes. Such a board no longer maps the process, but manages tasks. This is a departure from the actual Kanban principle. Here, the process is implicitly reflected in the tasks (Figure 34).

With task boards, the activities are displayed on the cards and not in the columns as with a process board. This means that a task board can be used flexibly for all kinds of activities and is not tied to a specific process. In return, however, there is no overview of the work progress from a process perspective. A simple task board only has four columns: Backlog, ToDo, Doing and Done. You are welcome to add further generic steps, such as a review, as additional columns, or set up an area in the Doing column where tasks are stored that are waiting for external input.

Each type of board, process board or task board, has its specific advantages and disadvantages. Only if the process is mapped on the board can you optimize the process in the true sense of Kanban. Process boards also provide a good overview of the current degree of completion or the current burndown from the backlog. With process boards, bottlenecks can be localized and buffered, and a flow can be created on the board via the WIP limit. The blockages created by the WIP limitation show potential for the next improvement steps. The disadvantage is that you cannot map tasks that follow a different process here. You can set up a process-free zone for these, which corresponds to the Doing column on a task board.

Task boards, on the other hand, are flexible and simple. All types of activities can be mapped. For example, different types of tasks can be displayed with different colors. However, a task board only gives you a limited overview of the flow in the process and only indirectly an impression of the current processing status.

Practical tip: Type of flow

The aim is always to create a flow and identify bottlenecks and blockages. Therefore, try to use process boards whenever possible. The resulting effort for defining the flow items and their granularity is important in order to make the way of working more flexible and to reduce the number of bottlenecks in order to lay the foundation for an agile way of working. I see the purpose of task boards above all when it is primarily about optimizing interfaces and not processes. By this I mean, on the one hand, the aim of protecting teams from the uncontrolled introduction of tasks by prioritizing and limiting tasks. On the other hand, task boards and their backlogs can be used to map collaboration between different organizational units in order to minimize latency times and make prioritization transparent.

Swim lanes

SSwim lanes are taken from the design of swimming pools. Just as every swimmer in a competition moves within a marked-off lane, you can divide your Kanban board with horizontal lines to separate different card flows from each other and to get a better overview (Figure 35). Whether the swim lanes start in the backlog or in a column further to the right depends on the purpose for which you want to use the swim lanes.

You can use swim lanes, for example, to separate different types of cards from each other, or different clients or different processing sub-teams. The WIP limits of a Kanban board apply across all swim lanes. Of course, you could also define a separate WIP limit for each swim lane, but in doing so you have actually created a separate Kanban board for each swim lane.

What you do with the swim lanes is up to you, there are no specifications in Kanban. However, it has become common practice to use

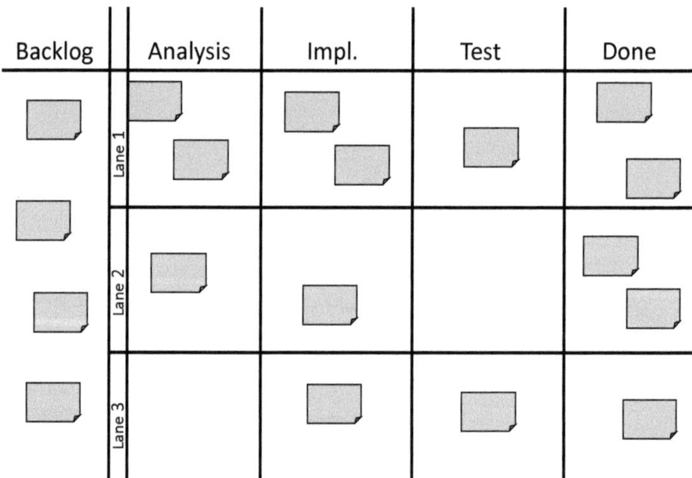

Figure 35: Board with swim lanes

    Kanban in Development

swim lanes to separate task types. Priorities, on the other hand, can be mapped using other methods, more on this in the Prioritization chapter. However, if it makes sense for you to model priorities with swim lanes, there is nothing that says you can't do so.

Estimates

When David Anderson experimented with his first boards and began to optimize his process, he found that the effort estimates for the maintenance tasks to be completed consumed up to a third of the total processing time. At the same time, in his context, the estimate had little impact on management's decision as to whether the maintenance task would be implemented at all. For Anderson, it was therefore obvious to do without the estimates altogether.

The basic approach with Kanban is to use throughput measurements to make statements about runtimes and not to estimate each task. However, the size of the tasks often varies considerably, so that it makes sense to divide the tasks into size categories. The task should be divided into size classes and the lead time measurements related to these size classes. To make the estimate independent of the level of experience and therefore the speed of the employees, you should evaluate the complexity and risk of the task in comparison to other tasks rather than the time required for the size class. Please also take a look at the Relative Estimates section in the Scrum chapter.

To evaluate the categories of complexity, it has proven useful to use clothing sizes such as S, M, and L. With the extended sizes XS and XL, you then have five categories available to describe the size class. For a common understanding, it is important here that you define a reference. This is a medium-sized, common or known task in the organization, which you define with M.

By working with the clothing sizes and corresponding measurements, you can quickly get a feel for which size has approximately which lead time in your process. If you make this transparent, this is important information for your stakeholders who schedule tasks.

Service classes

As mentioned above, the prioritization of cards is usually not mapped via swim lanes, but in service classes. The concept is similar to air travel. You can choose between different classes of service for a scheduled flight.

Certain rules are associated with service classes as to how the cards are moved across the board. Service classes therefore represent prioritizations. As with airlines, they therefore have different scopes of service and different prices.

David Anderson proposes three service classes, they should only be seen as examples. You may define service classes to suit your own environment.

- Standard: This card is subject to the WIP limits and is pulled according to the agreed pull rules.
- Accelerated (often also called Express or Silver Bullet): A card of this class is not subject to WIP limits and can be pulled immediately. Only one such card may be in the system at any one time.
- Fixed date: This card is subject to the WIP limits. The effort for this card is estimated, or at least a size category is defined. The status is monitored regularly. If the deadline appears to be in danger, the card can be transferred to the Accelerated class.

If you use different service classes, you must ensure that the stakeholders do not generally use classes with higher priorities such as Accelerated. There are basically two ways to do this: The simplest option is to allocate service classes. This can be done, for example, with the above-mentioned limitation on the board, or with the following rule: Stakeholders who submit tasks may create a maximum of three cards per month in the Accelerated service class. The other option is to narrow the comparison with airlines: If your organization is prepared to charge for services internally, you can also regulate the number of cards via the price. This is the elegant, but also more complex way of limiting cards with prioritized service classes.

Policies

As repeatedly stated, Kanban is merely a tool and not a framework. You have to define the way you work with your Kanban board in some way. This is usually done with policies, i.e., sets of rules. These define various aspects relating to your Kanban board.

In my opinion, the most important policy is the Next card policy or pull policy. This is a rule as to which card should be pulled next by the predecessor if there are several cards in Done. A good standard pull rule is: Pull the oldest card with the highest priority.

If you generally arrange the cards geometrically according to their priority and the most important card is always at the top, the simple and correct rule would be: Always pick the top card. However, this rule becomes softer the further the card moves to the right on the Kanban board. Especially with high WIP limits, i.e., many cards on the board, this rule makes it difficult to keep track.

A Definition of Ready can be used to determine when a card may be included in the Selected Backlog. If a card is allowed to be moved to a Done sub-column is defined in a Definition of Done that applies to this column. These two policies are explained in more detail in the Scrum chapter; you can also transfer the basic ideas described there to your Kanban board.

Another important regulation is the idle policy. What should employees do if the flow is blocked and they can no longer continue working? This policy helps to avoid uncertainty among the team and management. Common tasks in the idle policy include supporting the blocked position, further training or infrastructure work. Be sure to draw up this policy in collaboration with your team so that everyone has a common understanding of the tasks and supports them accordingly.

If things are unclear when dealing with the Kanban board, do not shy away from introducing additional policies. These should not be bureaucratic monsters, but be comprehended easily in just a few key words. Go to the team with your problem and work through the missing policies together with those who have to implement them. The system only works if it is supported by everyone.

Working with Kanban

WIP limits

In the section on lean in production and when introducing the Kanban boards, I already briefly touched on the topic of WIP limits. However, I limited myself to the primary effect: The reduction in throughput time. However, WIP limits also have a whole range of other positive effects.

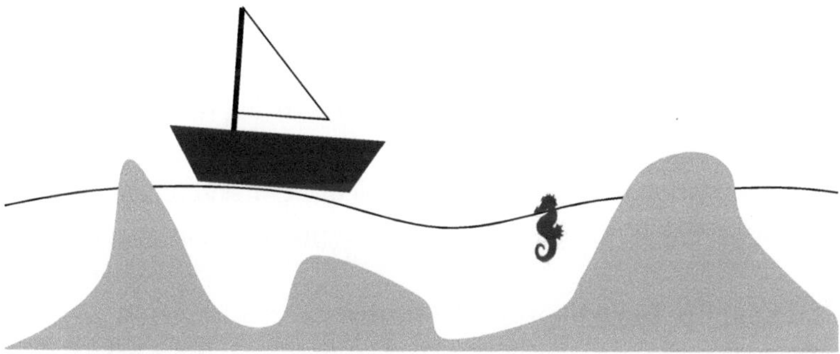

Figure 36: Problems become visible with a small WIP

The difficulty with implementation: WIP limits are initially counterintuitive for many, as the total amount of work to be done and the resources available for it remain constant. WIP limits then sound like less work and clients are afraid that their request will not be processed. In fact, WIP limits reduce the workload, but according to the queueing theory, this significantly shortens throughput times, especially in busy organizations. The supposed disadvantage that many tasks pile up visibly in the backlog (and not invisibly in the system as before) is an advantage with WIP-limited systems: Tasks that are still in the backlog can be changed or reprioritized at any time without affecting the development team. The advantages achieved by limiting WIP are therefore early feedback and possible late changes. Both help to reduce unnecessary work and reduce the effort required for revisions and changes.

As has become clear from the description of the Kanban concept, WIP limits also mean that the flow of tasks repeatedly comes to a standstill, that the system blocks when problems occur at a process step. This sounds dramatic at first. However, blockages are an important means of uncovering systemic problems in the process and optimizing the organization. A blockage uncovers problems that are often invisible in a clogged organization. Similar to a high-water level that hides dangerous rocks and reefs, high WIP makes the system's problems disappear (Figure 36). Without WIP limits, it is obvious that everyone involved will work past the existing problems instead of taking the time to eliminate them sustainably. A blockage forces the organization to deal with the problem. As I said in the production chapter: A blockage is the process improver's gold nugget.

Practical tip: Introducing WIP limits

Up to this point, I have described the effects of WIP limits in a steady-state system. If you are setting up a new Kanban board, it is best to start with high WIP limits so that no blockages occur at first. I do not recommend starting Kanban without WIP limits. This concept should be second nature to everyone right from the start.

Why not start with more aggressive limits right away? The people involved first have to get to grips with the board and the rules, i.e., the new way of working. Creating blockades leads to unnecessary uncertainty or even frustration at the beginning. As soon as the mechanism of the Kanban board has been established, start gradually lowering the WIP limit until the first blockages occur. Then start solving the underlying causal problems to re-establish a flow on the board. Then carefully lower the limit further until new blockages occur. Repeat this to further optimize the process and organization. At the same time, make sure that there are not too many blockages that could frustrate employees, or lead to no results at all. Working with Kanban boards means regularly rethinking and adjusting the WIP limits – both upwards and downwards. However, these adjustments must remain strategic in nature. Anyone who succumbs to the temptation to tweak the limits depending on the situation is torpedoing the Kanban system.

Blockages free up resources, which are then used for optimization in accordance with the agreements in the idle policy. If there is also idle time outside of the process, this is referred to as slack time. As motivation and innovation are important in your team, you should not define the idle policy too narrowly so that there is time for further training, basic research, and innovation. This is often disguised as a seemingly unproductive conversation in the coffee kitchen, but it moves the organization forward faster than employees working at maximum capacity.

It is not only the working atmosphere that influences the motivation of the teams, the short lead times themselves also have an important positive influence on this. Short cycle times offer a regular sense of progress whenever a small task is successfully completed. For me, employee motivation is the most underestimated factor in throughput optimization. Employee motivation can quickly lead to a double-digit percentage increase in throughput. This is often much more difficult to achieve with process changes. Motivating employees rarely requires large monetary investments. Nevertheless, systematic measurements and improvement programs for employee motivation are still not yet in place everywhere.

Another important aspect of WIP limiting in development is the impact on the human brain. The reduction of context switches caused by WIP limits has a twofold effect on the work of your developers. It reduces the frictional losses caused by mental setup costs and increases the quality of work. I write more about the losses caused by context switches in the People and Teams chapter.

Cumulative Flow Diagram

Cumulative Flow Diagrams (CFD) are the most important tool for evaluating and improving a Kanban implementation. The structure of a CFD is very simple. It is a stacked line diagram that visualizes the number of cards per process step over time. The last state (Done) is the lowest state in the diagram. A CFD therefore has the time, usually in days, on the horizontal axis and the number of cards in a particular state on the vertical axis. For the Kanban board in Figure 35, a CFD could look like Figure 37, for example. The wedge-shaped area at the bottom of the diagram is the Done state – it is constantly growing as more and more is completed.

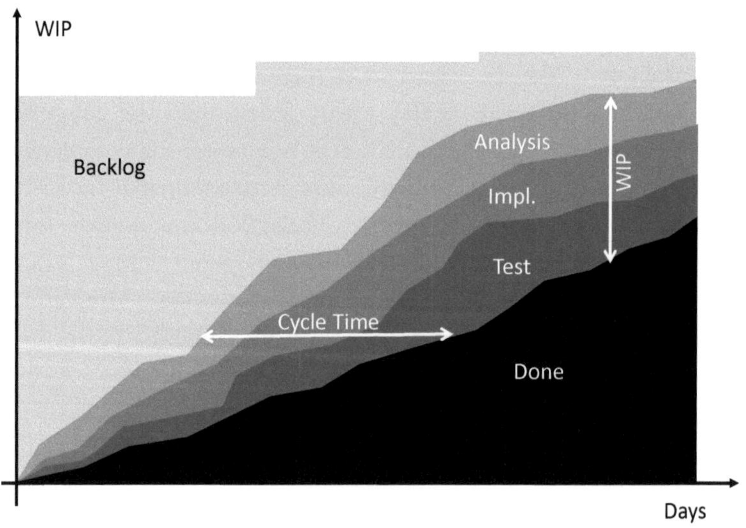

Figure 37: Cumulative Flow Diagram (CFD)

For any given day, you can simply cut vertically through the diagram and see how many cards were in each process step at that time, i.e., how high the WIP was. If, on the other hand, you aim from a transition point from the backlog in Analysis to the right up to the edge of the Done area, you can read off the cycle time at this point in time. By interpreting the curve shapes, you can read a great deal of information from this simple diagram. In addition to the previously WIP and cycle time you can also interpret

- Blockages that have occurred (flatlines)
- Overcapacity (conditions with too small of an area)
- Problems with the pull system (states with too large of an area)
- Processing in batches (steps)
- Quality risks (flanks too steep, tickets moved under pressure)

Cadence

If you know Scrum, you know that after each Sprint there is a new version of the product, and by definition this should also be potentially usable. Kanban does not initially define any release cycles or timeboxes; the user must do this themselves. Kanban initially creates a continuous flow via the Kanban board. However, as you probably don't want to use Kanban to develop with a waterfall model, you need to define a regular interval in which the processed tasks lead to a potentially usable product. In flow-oriented systems such as Kanban, this interval is not referred to as a sprint, but more abstractly as a cadence.

Introducing a cadence for Kanban boards is important for various reasons. If you use cadences to provide tested product increments in a similar way to Scrum, you can receive important feedback from your stakeholders regularly and at an early stage. This uncovers misunderstandings in the implementation of the specification and creates opportunities for new ideas and innovations. Cadences therefore help to avoid unnecessary work and make the product better than the specification actually envisages. Another important aspect is that cadences catch hidden drifting project attributes. These are, for example estimates, progress measurements, and quality attributes. In other words: After each cadence, you know where you are at, and a new small project section with a new plan is created. You do not necessarily have to build product increments in time with the cadences. Without products, however, it is much more difficult to reliably determine the status quo of development. When it comes to estimates in particular, many people believe that the sum of the estimation errors or chance will somehow cancel each other out. The reality, however, is that chance drifts. Take the time to flip an ideal coin 1000 times (at least in your mind). Starting at zero, add the value of one for each emblem and subtract the value of one for each number on the coin. You will find that the added-up results of the random tosses do not stay around zero as assumed, but drift. Such coin toss attempts look something like Figure 38.

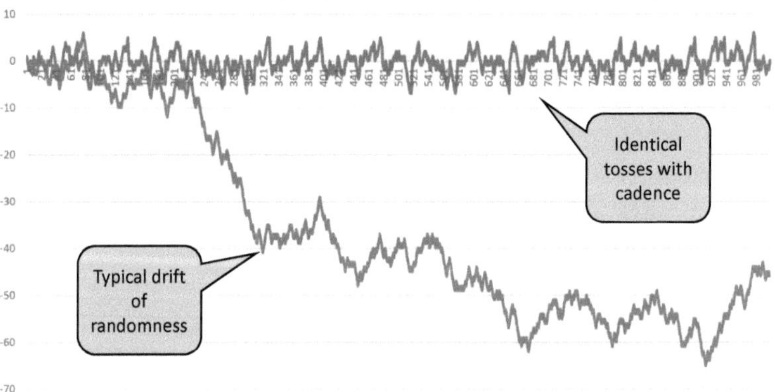

Figure 38: Cumulated coin tosses with and without cadence

 Kanban in Development

Note: Each toss of the coin has a 50:50 probability of occurring again and therefore cannot actively return to the zero line. Cadences interrupt such drifts of the project variables because they settle accounts with the previous phase and initiate a new phase. As mentioned, this applies not only to estimates, but to all project attributes. The illustrated coin toss with a cadence of 20 tosses is shown in Figure 38 with the second line, the drift is visibly reduced.

If you are thinking about what cadence makes sense for your Kanban system, the basic rule is the shorter, the better. Or put more precisely: as short as possible and as economically viable as possible. Given that the definition of the cadence must also be economical. It depends on your transaction costs, i.e., the time and effort required to create a cadence from the data and information of the current development process to create and test a (partially) functioning product. These transaction costs vary significantly depending on the product, technology, and industry. Please also read the section Batch Sizes in the Production and Lean chapter.

Kanban roles

Unlike Scrum, Kanban is not a framework, just a tool. Kanban therefore does not define any roles or responsibilities that are useful for working with the board. In addition to assigning employees to the individual process steps on the board, you need a person responsible for the backlog. This person should ensure the quality of the content and also be responsible for prioritization. You can introduce the role of the Product Owner here, similar to Scrum, or form a committee that is responsible for filling and prioritizing the backlog. Personally, I would always prefer a single, coordinating and decisive person over a committee. Millions of Scrum implementations have shown that this is a very powerful setup. Ultimately, however, how this responsibility is fulfilled is a question of taste.

In addition to responsibility for content, you should define who is responsible for complying to the rules of the game, i.e., policies, and actively supports problem solving in the event of blockages. Given that the development team focuses on product development, it makes sense – in my opinion – to outsource these tasks to a separate role analogous to Scrum, which takes care of the team and obstacles in the organization. The Scrum approach has been tried and tested for 20 years, so there is nothing to stop us from drawing on this experience and introducing the role of a Kanban Master. This person takes care of the process, the team and, above all, the incremental improvement of the organization.

First rule of scaling agile: DON'T!
Andreas Rowell

Scaling Scrum

Introduction

A team of teams

In this context, the term scaling refers to the expansion of agile working methods to several teams or even several products. If you leave the one team, one product context of Scrum, you need to think about coordinating multiple teams in Scrum mode. This usually involves

- Concepts for planning a cadence across teams
- Concepts for coordinating different backlogs
- Concepts for coordinating different roles
- Concepts for coordinating the collaboration of several teams

All agile scaling concepts dispense with a centrally managed approach: Just as the members of a team coordinate and work together in a self-organized manner, collaboration between several teams should also be coordinated directly – i.e., without a central management entity. A single team becomes a team of teams.

Various frameworks and concepts have emerged for scaling agile teams that promise an agile way of working beyond the context of a single team. The established frameworks certainly focus on different organizational sizes, even if many elements of the working methods of the individual concepts are very similar. The mechanisms of Scrum remain the same.

In practice, the challenge lies less in the direct implementation of scaling concepts, but rather in the existing framework conditions in terms of corporate culture, skill sets, organizational structures and locations, as well as system architectures.

Basics

Functioning agile teams form the basis of every scaling process. What changes at team level is that the team's Product Owner is no longer responsible for the entire product, but only a part of it. In addition, the Developers cannot usually build and test the product completely autonomously at the end of the sprint, but must work together with other teams.

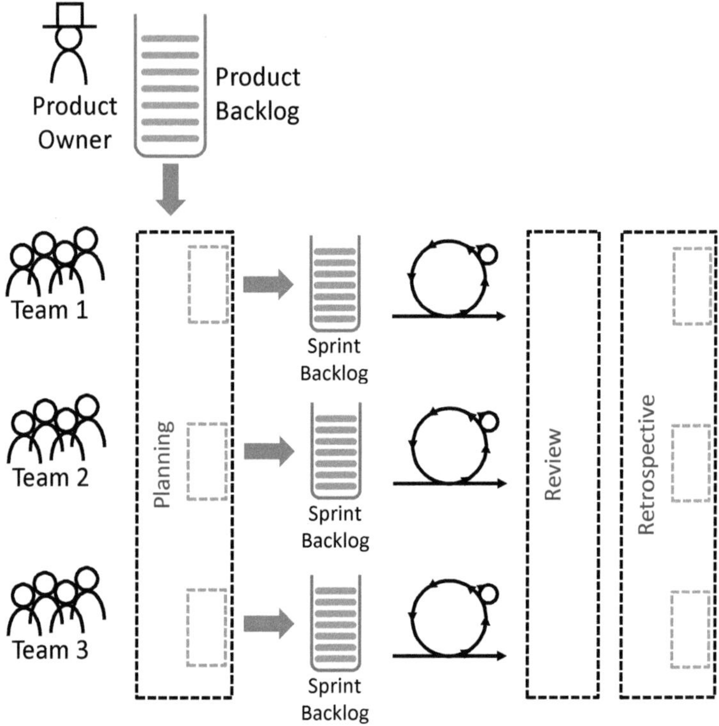

Figure 39: Scaling principle

In any case, the Product Backlog remains product-related, so there is still only one Product Backlog from which various team members can pull. There is also a top-level Product Owner who has an overview of the product and is responsible for communicating with the stakeholders. As a rule, scaling therefore requires that both the

Product Owner and the backlog are structured in a kind of hierarchy, with different levels for Product Owner and backlogs. The different backlog levels also have different levels of granularity, as the Product Owner can no longer manage the backlog items for several teams with the granularity that is usual for Scrum.

With scaled agility, the principles of bottom-up planning are retained. This means that the teams are still involved in the planning of cadences. The planning meetings take place synchronously, ideally with all participants in one room, so that the teams can also identify mutual dependencies during the planning process and manage them in a self-organized manner. The teams then work through their own backlog in order to integrate and test a product increment with the other teams at the end. This is the task of the Developers; under no circumstances is a dedicated integration and test team deployed.

Figure 39 shows an abstraction of the structure of backlogs and Product Owner. The Scrum Master structure is usually scaled in parallel. Impediments that cannot be resolved by Scrum Masters at team level are passed to the next higher Scrum Master level. However, I have not included this aspect in the diagram for the sake of clarity.

Descaling

The concepts described assume that stable, experienced Scrum teams are scaled. At the very least, experience shows that scaling only works properly in this case. The great danger is that existing dysfunctionalities of the organization is scaled up via frameworks – and this naturally does not lead to the desired success. Becoming agile means dissolving dysfunctional aspects. As Jeff Sutherland says so well, "Crap does not scale."

Here some problems that are often scaled instead of solved:
- Contracting of team members to the project <100%
- Spatially distributed development, often across continents
- Retention of project and matrix structures

- Unfavourable team structure
- External team dependencies
- Inflexible system architectures
- Insufficient test and build automation
- The Scrum Masters are not taken seriously or there is no willingness to change

All of these issues should actually be solved for this team with a stable Scrum implementation in just one team. Suppose, based on your experience with conventional development, you expect to have five development teams and want to scale Scrum using an appropriate framework. If you eliminate the aforementioned dysfunctionalities with just one team, this team will most likely become so productive that you can deploy the other four teams for other tasks.

Every additional person in the project creates additional overhead in communication and coordination. Your goal should be to develop your product with as few people as possible – only then will your organization be flexible and fast; only then will it be agile. Resist the reflex to deploy even more developers on the project if the project is running too slowly. This will only make the project even slower. Work on the dysfunctionalities, i.e., the impediments, in order to become faster. It is often even advisable to reduce the size of the development team in order to become faster. This concept is known as descaling.

The assumption that development performance increases linearly with the number of people involved is widespread. A linear relationship exists, if at all, for mechanical work, but not for creative collaboration. Take this example: 20 people are not able to sort a stack of playing cards four times faster than five people. An interesting example of this is the FBI Sentinel Project, an IT project for the management of FBI files (Sutherland, 2014). After the US government had spent several hundred million dollars and received only a fraction of the desired functionality, the project was stopped and restarted. The project team was reduced from 400 to 45 people. After 12 months, the software was finished at only ten per cent of the estimated cost.

Available frameworks

In the following sections, I will present four widely used frameworks for scaling agile approaches:
- Nexus
- Large Scale Scrum (LeSS)
- Scaled Agile Framework (SAFe®)
- Scrum@Scale

The definitions of these four frameworks are available on the Internet for free. For this reason, I have refrained from using graphical representations of the frameworks in this book and instead provide the Internet address with the definition for each framework.

Nexus

The Nexus framework by Ken Schwaber or scrum.org is the most recent and leanest of the frameworks presented here for scaling Scrum. The first Nexus Guide was published in 2015, the current version can be found at www.scrum.org/resources/nexus-guide. Nexus adheres closely to the Scrum Guide and accordingly sets the size of the Nexus, the team of teams, at three to nine Scrum Teams.

The framework redefines the Scrum events at Nexus level and specifies the interaction between Nexus and team level. It also defines a new role: the Nexus Integration Team.

As with Scrum, the starting point is a Product Backlog for which a Product Owner is accountable. This person is also responsible for the refinement of the Product Backlog, in which the entries are broken down into smaller units and the teams expected to implement them and the dependencies between the work packages are identified. In contrast to Scrum, the refinement is an official event and is carried out by the Product Owner together with representatives from the Scrum teams.

The Sprint begins with Nexus Sprint Planning, in which team representatives take the high-priority items from the Product Backlog and bring them to their teams. In the second step, each team carries out Sprint Planning in accordance with the Scrum Guide. The teams coordinate according to the dependencies identified previously or in the Sprint Planning. The selected Product Backlog items (PBIs) and their dependencies are visualized by the teams in a matrix in the Nexus Sprint Backlog.

Development work begins after the teams' Sprint Planning. The product is continuously integrated by all teams involved in order to identify problems at an early stage. The Nexus Integration Team (NIT) – a separate role newly introduced by Nexus – has become important. This team is accountable for the integration of the product, even if the integration is carried out operationally by the Scrum Teams. The Nexus Integration Team consists of experienced developers who can advise Scrum Teams on test and build automation and, if necessary, system architects or other people who can act as technical coordinators and consultants for the teams. The members of this team can also be members of the Scrum Teams. In this case, it should be noted that the work in the Nexus Integration Team always takes precedence over the work in the individual teams. The Product Owner in the Nexus Integration Team is also the Product Owner of the product and is accountable for it. The Scrum Master of this team is responsible for the implementation of the Nexus framework.

The Nexus Daily Scrum, a coordination and planning meeting with team representatives, takes place every day. At this Daily Scrum, the progress of the product is inspected and integration problems and new dependencies are identified. With the results of the Nexus Daily Scrum, each team starts its own Daily Scrum at team level.

At the end of the sprint, all members of all teams meet for the Nexus Sprint Review. This event is the only one in Nexus where no representatives are used – everyone should be present. Nexus does not have reviews in the individual teams, as the focus is always on the product as a whole. In contrast, the retrospective again takes place at both levels, team, and Nexus. The retrospectives at team level are framed by the Nexus. A Nexus Sprint Retrospective therefore takes place initially, in which team representatives identify potential for improvement. Using these results, each Scrum team then conducts its own Sprint Retrospective. In the third step, team representatives present the results of the team level in a second part of the Nexus Sprint Retrospective. This is where the participants decide how to visualize and track the identified measures.

Nexus focuses on smaller scaling environments. It comes from software development, but is also used in the development of mechatronic products. If you are familiar with Scrum, you will also be able to quickly become productive with Nexus, as the framework is so closely based on teamwork with Scrum.

Large Scale Scrum (LeSS)

From 2005, Craig Larman, one of the co-signatories of the Agile Manifesto, developed a model together with Bas Vodde for implementing Scrum with several teams. The experience gained with some customers resulted in the model that is known today as Large Scale Scrum. The definition of the framework and additional information can be found on the Internet at less.works.

The core mechanisms of the LeSS framework are very similar to the Nexus approach, or rather Nexus is very similar to LeSS, as LeSS was created around ten years earlier. Both approaches are very clearly oriented towards Scrum and scale the accountabilities, events and artifacts – hence the great similarity. In contrast to Nexus, however, LeSS provides a comprehensive superstructure of principles, experiments, and guides. This provides users with a pool of information that supports the implementation of LeSS – because there will also be no best practices.

Principles

The principles of the LeSS model are the basis for all decisions on how LeSS should be implemented in a specific context. Here is an overview of the ten principles – find out more on the LeSS website:

- Large Scale Scrum is Scrum
- Transparency
- More with less
- Whole-product focus
- Customer-centric
- Continuous improvement towards perfection
- Lean thinking
- Systems thinking
- Empirical process control
- Queuing theory

Framework (the rules)

The framework, which is based on these principles, specifies the rules for how Large-Scale Scrum should be executed.

As with Scrum, there is a Product Owner who maintains a Product Backlog. Implementation is carried out by two to eight teams, which are ideally set up in a cross-functional way so that they can work independently on as many types of tasks from the Product Backlog as possible. As with Scrum, Scrum Masters work at team level. The Product Owner therefore only exists once – across teams – while there may be multiple Scrum Masters at team level (according to LeSS, a Scrum Master can support one to three teams).

The three artifacts Product Backlog, Sprint Backlog, Increment, do not differ from Scrum, except that the Sprint Backlog exists several times, one for each team.

The Scrum events are scaled by holding them at team level and at the team of teams level. In Sprint Planning One, all teams or team representatives meet with the Product Owner. The Product Owner presents the top items in the Product Backlog; the team representatives then divide these up among themselves and also discuss what needs to be coordinated between different teams in the next sprint. The representatives then go back to their teams and start Sprint Planning Two. In this event, the teams plan the details of the current Sprint and coordinate with other teams where there are dependencies in this sprint. The Daily Scrum takes place at team level. A corresponding meeting across teams is not mandatory and takes place as required. All teams meet for the Sprint Review with the Product Owner and, if necessary, with stakeholders. As it is about the integrated product, it makes no sense to conduct a review at team level. For the retrospective, on the other hand, LeSS relies on the division into two parts: First, each team carries out a retrospective, followed by an overall retrospective, in which the team representatives, Scrum Masters, Managers, and Product Owner meet to define improvement measures across the teams.

Experiments and Guides

The creators of LeSS provide the user with additional guides. These include some 100 articles on the LeSS website with tips and suggestions for practical implementation. These guides are grouped into the following subject areas:

- Structure
- Management
- Technical excellence
- Introduction of LeSS

Larman and Bodde point out that their first two books on scaled Scrum (Larman & Vodde, 2008, 2010) mention a large number of experiments that have worked well for scaling in various organizations. Experimentation is an essential part of LeSS as it allows an organization to find the path to scaled agility that works for them.

LeSS Huge

In organizations with more than eight teams, the Product Owner in LeSS would become a bottleneck. For such organizations, Larman and Vodde have defined a second framework, the LeSS Huge. The main difference is that the requirements in the Product Backlog are assigned to Requirement Areas. This creates a two-stage Product Backlog: The actual Product Backlog and several Area Product Backlogs, which are managed by corresponding Area Product Owners (APO). Each Area is essentially a LeSS implementation with four to eight teams. LeSS Huge also results in a single integrated product at the end of the Sprint, into which the work of all teams has been incorporated. Numerous guides are also available for LeSS Huge, which make it easier to get started in such a large environment. The LeSS Huge documentation rightly points out that LeSS Huge cannot be introduced with a big bang. Users should be prepared for the fact that it will take years before such a large number of people can work together in an agile way (this applies to all scaling frameworks, by the way).

Scaled Agile Framework (SAFe®)

SAFe stands for Scaled Agile Framework, the brains behind this framework is Dean Leffingwell. In 2010, Dean incorporated various agile concepts as well as the ideas of Donald Reinertsen, and he presented a possible process model for scaled agile development in a large diagram (the Big Picture) in his book Agile Software Requirements. This clear diagram met with great response from potential users. Dean and a small team then further developed his ideas into a framework for scaling, and published it in version 1.0 as the Scaled Agile Framework in 2011. The framework then evolved through feedback from users in several stages up to the current version 6 (2024), which I describe here. You can find the description of the current version on the Internet at www.scaledagileframework.com

SAFe defines a work structure through four different Configurations that can map different organizational levels. SAFe is the only framework listed here that explicitly addresses topics such as multi-product environments, portfolio management or documentation, and compliance. The creators behind SAFe have proposed concepts for many aspects of system development, which is why the scope of the description differs significantly from that of the other frameworks. While the Nexus Guide comprises around 15 pages, the definition of SAFe was available as a book in version 4.6 with over 850 pages.

In the following, I will briefly describe the core concepts of SAFe with all four possible configurations.

Essential SAFe

The simplest configuration in SAFe comprises the team level with several teams and the level of a team of teams. The teams do not necessarily have to work with Scrum; other agile working methods are also possible as long as they include the four elements of planning, implementation, review, and retrospective. These four steps are carried out in two-week iterations. SAFe chooses this term in order to be method-neutral; when using Scrum, the iterations correspond to Sprints. Backlog items at team level are referred to as Stories. These must be small enough to fit into the two-week iteration.

Every agile team consists of the three (former) Scrum roles: A Product Owner who manages the content for the respective team, a Scrum Master who looks after the team and is responsible for improvement and speed, and Developers who have technical responsibility.

The Team of Teams level is called ART (Agile Release Train) in SAFe. Accordingly, there is an ART Backlog at this level instead of a Product Backlog, which describes the Features – the functionalities at this level. An ART Backlog can also contain features for different products. The team level and the ART level are firmly connected and cannot be used separately, as the ART provides the teams with work. This configuration, consisting of the two lowest levels, is therefore the smallest of the four possible SAFe configurations.

With the scaling of the Scrum Product Backlog to the ART Backlog, SAFe also introduces a scaled cadence: SAFe places a second interval, the Planning Interval (PI), over the two-week iteration. The cadence of the PIs can be set by the organization between eight to twelve weeks, which corresponds to four to six iterations. A PI is merely a planning period. In contrast to a Scrum sprint, a PI does not necessarily result in an Increment at the end, as this would be too inflexible with a cadence of two to three months. Releases – i.e., tested product increments – are created as required. The slogan for this is: Develop on Cadence – Release on Demand. In line with the longer cadence of a PI, Features (the backlog items in the ART Backlog) also have a different size. They can span iterations, but must still be small enough to fit into a PI. Features and Stories are in a hierarchical relationship, and a feature is broken down into 1-n stories.

Regarding the Team of Teams, an ART should consist of no more than 150 people. This is roughly the maximum size of a group in which the members know each other, work together, and can trust one another.

The agile team roles with the responsibilities of content, process and technology scale to the following roles at ART level:

- The Scrum Master at this level is the Release Train Engineer, or RTE for short. They take care of the process and the elimination of impediments at this level.
- As there can be several products in the ART backlog, there is no longer a Product Owner at this level, but the role of Product Management, which can also be filled by a small committee.
- The technical expertise of the Developers scales at the ART level to a group of system architects or system engineers who define the technical guidelines for all teams.

Planning concept

What makes SAFe special is the planning process based on the essential configuration, known as PI Planning. This is usually a two-day planning meeting with all those involved, which in practice can mean that 150 to 200 people meet in one room. In this meeting, the content for the upcoming PI is planned and dependencies between the teams and planning risks are identified.

The first morning of PI Planning is dedicated to managers and architects. Presentations on the market, product, vision, and architecture create a common understanding of the current situation and the technical framework conditions for the coming months are defined. The top 10 Features in the program backlog are also presented to all attendees one more time.

Then the core of PI planning begins, in a first team breakout. The teams pull Feature by Feature from the ART Backlog in a self-organized manner, break them down into individual Stories and identify and manage dependencies between the teams. As the other teams are only a few steps away, this is a simple exercise. As a result of the coordination, further stories are created, which each team plans in addition to their existing stories and physically distributes as cards at their team table for the upcoming iterations. This does not replace the planning of iterations every two weeks; it is merely a preview of the possible workload and the current dependencies in the PI. When managing dependencies, the priority of the features is inherited by that of the derived stories. There are therefore no conflicts as to which feature a team should work on first with other teams. When a dependency is identified, copies of the interdependent backlog items are created and attached to the so called ART Planning Board. All teams and iterations are shown on the board, and the dependencies themselves are visualized by a thread between the items. This supports the planning process and helps to manage dependencies during the PI. This visualization also allows unfavourable team assignments to be identified and adjusted if necessary. In this planning process, only 50 to 70 per cent of the team velocities are planned, as the ART backlog only contains the currently known topics. There should therefore be room for those topics that arise during a PI.

After the first team breakout, the existing draft plan is discussed by everyone in the room in order to gain a general understanding. The stakeholders in the room may recognize a deviation from their own ideas – as with all agile approaches, they have to accept the natural speed of the organization. However, this can give rise to new ideas on prioritization and the layout of the features, which are discussed and agreed by the stakeholders on the evening of the first day.

The second team breakout takes place in the morning of the second day after the management decisions from the previous evening have been announced. With this information, the teams now begin to finalize the PIs' planning. At the same time, the teams identify the risks behind the planning on cards or sticky notes, for example: lack of supplies, absence of experts, unclear requirements, questionable technical feasibility.

After a joint review of the final plan, the risks are discussed and categorized in a plenary session. Through this exchange, all participants gain a common understanding of the planning and the risks for the PI. In a confidence vote, each team member then signals by a show of hands whether they have confidence in the plan. This helps to to first emphasize that it really is a pull system and that the teams themselves can decide what fits into a PI and what does not. Secondly, the aim is to identify technical experts who may have been overlooked. Their opinion should be heard so that the plan can be adapted again if necessary. However, this rarely happens in practice.

Other configurations

The levels above the Essential Configuration are optional and can be used independently of each other. As the actual mechanism of SAFe lies on the Essential Configuration, I will only briefly discuss the two upper levels. The Large Solution level can be implemented above the ART level. This is necessary if several ARTs are developing a product, as it provides mechanisms for synchronization between ARTs. The items in the Solution Backlog are referred to as Capabilities. A Capability is then broken down into one or more Features at ART level. In terms of size, however, Capabilities must be implemented in the same way as features within a PI. Just as the ART Backlog is scaled upwards, the three agile roles are also scaled from the ART level to the Large Solution level, where they are called Solution Management for the content, Solution Train Engineer for the process and Solution Architect for the technical view. At the Large Solution level, SAFe also offers concepts for dealing with specifications and compliance issues. The Large Solution configuration consists of the Essential configuration and the Large Solution level.

The top level, which can be added to the lower levels with or without the Large Solution level, is the Portfolio level with the corresponding Portfolio Backlog. At this level – company-wide or across business units – strategic goals are administered initiatives and called Portfolio Epics. In contrast to Capabilities and Features, these can also span several PIs. Depending on the configuration, they then break down into Capabilities at the Large Solution level or Features at the ART level. The Portfolio level can be combined as a Portfolio Configuration with the Essential Configuration or as a Full Configuration with Essential and Large Solution.

SAFe works with a Kanban system on each of the three upper levels to allow backlog items to mature until they end up in the backlog. The WIP limit restricts the number of simultaneous initiatives/Capabilities/Features. This concept is particularly interesting at portfolio level, as it establishes a pull system from top management to the Developers. In other words, new strategic initiatives may only be pulled into the Portfolio Backlog by management once the teams have completed another initiative, i.e., capacity has been freed in the organization.

The complete documentation for SAFe can be found on the scaledagileframework.com website behind the Big Picture, where all icons can be clicked for further explanations.

.

Scrum@Scale

Scrum@Scale is the framework developed by Scrum inventor Jeff Sutherland for scaling Scrum (www.scrumatscale.com). It differs from the three approaches presented so far in that it does not specify any fixed levels or structures, but instead only provides a scaling mechanism for two Scrum Teams that scales fractally to any size. While the three models mentioned above think in terms of fixed levels and units, Scrum@Scale creates a precisely fitting tree-like structure that also reflects the strict separation of the What and the How defined in Scrum. The basis for Scrum@Scale is – unsurprisingly – a functioning Scrum Team. While other models are often brought to life with a big bang introduction, Scrum@Scale starts with a single Scrum Team. Only when this is running well and the environment and management have really understood Scrum another team is added and coordinated using the mechanisms of Scrum@Scale. In this way, a structure of mature Scrum teams grows over time, ensuring that only well-functioning Scrum is scaled and that the existing dysfunctionalities of the organization are not expanded. This risk is definitely present in big bang implementations.

Scrum Master cycle

The Scrum Master Cycle takes care of the How at Scrum@Scale. The basis is the work of the Scrum Master at team level. The coordination of the How across several teams takes place within the so-called Scrum of Scrum (SoS). This is a Scrum team that is responsible for integrating the work products of the coordinated teams. However, this does not mean that this team carries out the integration itself. The SoS includes the three Scrum accountabilities from the Scrum Guide. In addition to the Scrum Masters of the coordinated teams, there are usually other people with the necessary expertise in this team, such as architects, quality managers, and so forth.

The Daily Scrum of the SoS is used to resolve impediments that could not be resolved one level below this level. A coordination team can be set up above the SoS to coordinate this Scrum of Scrums and others. This is then referred to as Scrum-Of-Scrum-Of-Scrums (So-SoS). This team is also responsible for the unresolved impediments of the level below and for coordinating the SoS. This creates a fractal tree structure in which impediments move upwards and in which the levels below are coordinated at each higher level. As a result of this and the change in the organization, the framework conditions are gradually optimized at each level.

This makes it clear that the management must be at the top of this Scrum Master organization. In fact, the top level is the Executive Action Team (EAT) – it is the SoS for the entire organization. The Executive Board or the Managing Director are therefore the organization's top Scrum Masters. All impediments that could not be resolved by the Scrum Masters at the levels below end up with the EAT – on a daily basis. This clearly emphasizes the claim of Scrum – especially the Scrum Master: Obstacles are cleared out of the way and work is constantly being done to improve the organization. In concrete terms, it is mostly about making the previous product life cycle and the associated processes more agile step by step, or as Jeff Sutherland describes it, creating an operating system in which Scrum can be executed.

Product Owner cycle

Scrum@Scale also brings the separation of the What and the How to the scaled organization. Consequently, when Scrum@Scale is introduced, a Product Owner organization is created in parallel to the Scrum Master organization described above, which coordinates the What across various levels.

To this end, the Product Owners of several Scrum Teams form a Meta Scrum – a Scrum Team that takes care of the backlogs of the teams to be coordinated. To this end, the Meta Scrum maintains its own backlog, which is then broken down into the backlogs of the underlying teams. The Product Owner of the Meta Scrum is the Chief Product Owner (CPO). This structure also forms a fractal tree, so that several Meta Scrums are coordinated by a Meta Scrum above it with a CCPO. At the top is the Executive Meta Scrum (EMS), which maintains the organization's top backlog at management level. This backlog usually includes strategic initiatives.

While the Scaled Daily Scrum is the main event from the perspective of the How in the Scrum Master organization, it is a scaled backlog refinement in Meta Scrum. The Product Owner organization therefore takes care of developing the strategy and vision, prioritizing the backlogs, refining and breaking down the backlog items and planning the time, budget, and release.

Connection of the cycles

A key point of contact between the Scrum Master organization and the Product Owner organization is collaboration at team level. This collaboration between the three Scrum accountabilities is described in great detail in the Scrum Guide. The second important point of contact is feedback on the product. When customers work with products, the development organization can review its own assumptions and constantly improve both the product (What) and the processes of development, construction, testing, and delivery (How).

Overview of the four frameworks

	Nexus	**LeSS**	**SAFe®**	**S@S**
Product Owner (Team)	yes	no	yes	yes
Scrum Master (Team)	yes	yes	yes	yes
Product Backlog (Team)	no	no	no	yes
Product Owner (scaled)	PO of the NIT	Product Owner	Product Mgmt.	CPO / EMS
Scrum Master (scaled)	S/M of the NIT	-	RTE	SoSM / EAT
Accountable for integration	NIT	Teams	Teams	SoS

Take people as they are,
there are no others.
Konrad Adenauer

People and Teams

People

Motivation

What drives people, what slows them down? To investigate this question, the US author Daniel Pink summarized various scientific studies in his bestseller *Drive – The Surprising Truth About What Motivates Us* (Pink, 2009). The works Pink found all come to similar conclusions – which, however, still have a long way to go in corporate management cultures.

The bottom line: While higher pay for physical work definitely leads to higher performance, this is not necessarily the case for creative work. In fact, a higher salary often leads to lower performance, especially if employees' expectations are disappointed. These effects occur from a salary level at which employees no longer have to think about money because their financial livelihood is secure. From this point onwards, other incentives are needed to motivate people.

While external incentives such as money are referred to as extrinsic motivation, the inner drive is referred to as intrinsic motivation. According to the studies in Pink's book, this intrinsic motivation is the only functioning drive in creative activities. According to Pink, this motivation consists of three pillars: autonomy, mastery and purpose. Intrinsic motivation is only possible if all three pillars are in place (Figure 40).

A good example of these three aspects is the motivation of open-source programmers. According to prevailing management thinking, it should not be logical for well-paid software developers to develop professional software in their spare time without receiving a penny for it. However, if you look at the commitment of open-source developers in relation to the three pillars of motivation, it makes perfect sense. Nobody is asking these experts to organize their free time in this way. They act voluntarily and can determine the scope and organization of their working hours themselves: Autonomy. They become experts in their field, and gain experience the recognition of the community: Mastery. The meaning of their work is tangible for them, millions of people use their software: Purpose. The three pillars can also be recognized, for example, in voluntary work and probably also in the work you do.

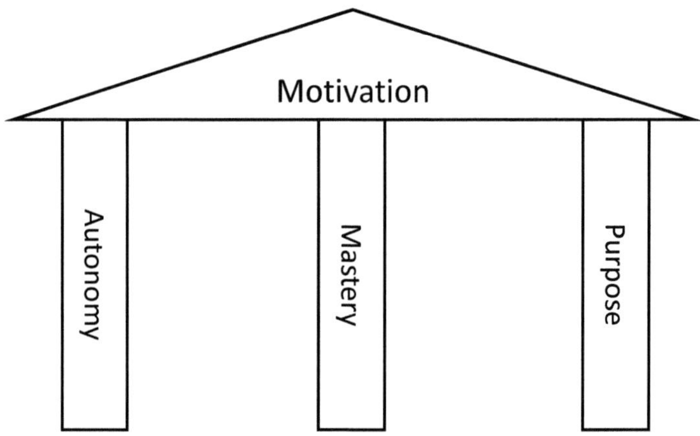

Figure 40: Pillars of motivation

If you correlate the three pillars of motivation according to Pink to the Scrum framework, you can also see how Scrum promotes the motivation of those involved. Scrum relies on the self-determined work of the three Scrum Accountabilities and the entire organization. The three Scrum Accountabilities act as peers, no one can give instructions to the others – that is autonomy. Through cyclical feedback in the Sprint retrospective and a protected space around the Scrum Team, Scrum enables a learning mini-organization and supports mastery in the workflow, and in the professional expertise of

                                        People and Teams

the individual in the self-organized team. The purpose of the work is much more tangible for Scrum Teams than in traditional settings, where many levels are often drawn in between the developer and the customer or their business, thanks to the close contact with the stakeholders and the product vision.

Games and invitations

Jane McGonigal provides a different view of motivation and drive in her book *Reality is Broken* (McGonigal, 2011). She came to the conclusion that people are motivated when their activity is designed like a game and therefore has the following four attributes:

1. A clear goal
2. Clear rules
3. Feedback on progress/score
4. Opt-in participation

For example, the aim is to score more goals than the opposing team in soccer or to get a ball into 18 holes in golf. Without the rules of the game, such as the offside rule, the ban on handball or the requirement to hit the golf ball to the hole from a great distance with a club, the task would be too easy. There would then be no challenge and no opportunity for the players to improve. The current score and also the match result are always visible to all participants in games and support the desire to win or strive for improvement from game to game. However, the high motivation to participate in a game with full commitment requires that participation is voluntary. This last aspect often causes difficulties for the management in companies, although it is the basis for commitment.

I would also like to make the connection to Scrum from this point of view: Scrum adds play to reality. This intention is even reflected in the subtitle of the Scrum Guide: The rules of the game. The aim of Scrum is to deliver a Done increment after each Sprint. The rules are written down in the Scrum Guide and are monitored by the

Scrum Master. All members of the Scrum team receive continuous feedback – from the stakeholders, through the burndown chart and the velocity measurements. This explains the enormous drive that Scrum Teams can develop – provided they are not forced by the organization to work with Scrum. You will certainly recognize parallels to Daniel Pink's three pillars of motivation here, they are just different perspectives on the same entity: People.

It is interesting to note that a good invitation consists of the same four attributes as a good game. Use this checklist if you want to win someone over for something – whether professionally or privately. Unfortunately, real invitations are very rare in our organizational cultures. Even if we often talk about invitations in everyday life, they are usually a summoning. If you don't accept the invitation, you are forced to explain. Try to focus more on genuine invitations. If no one accepts your invitation, either the invitation is not ideal or the invitees do not consider your plan to be worthwhile. Both give you the opportunity to improve something. Real invitations are therefore an important step towards more transparency and an opportunity to constantly improve.

Remember: Transparency, Inspection, Adaptation.

Context switching

In the early 1990s, Gerald Weinberg's book Software Quality Management argues that for every task we work on at the same time as a primary task, 20 per cent of our mental performance is lost due to the change of context – through mental set-up costs, so to speak (Weinberg, 1992). So, if you are working on three projects at the same time, each project will not have 33 per cent of your creativity and performance at its disposal, as initially assumed, but only 20 per cent. The rest is consumed by mental set-up costs, approximately 20 per cent per additional task (Figure 41). Context changes have an enormous impact on the performance of people and teams and therefore on the time and costs of your project. Nevertheless, in many organizations context switching is accepted without reflection, or their detrimental effect is not recognized or simply ignored.

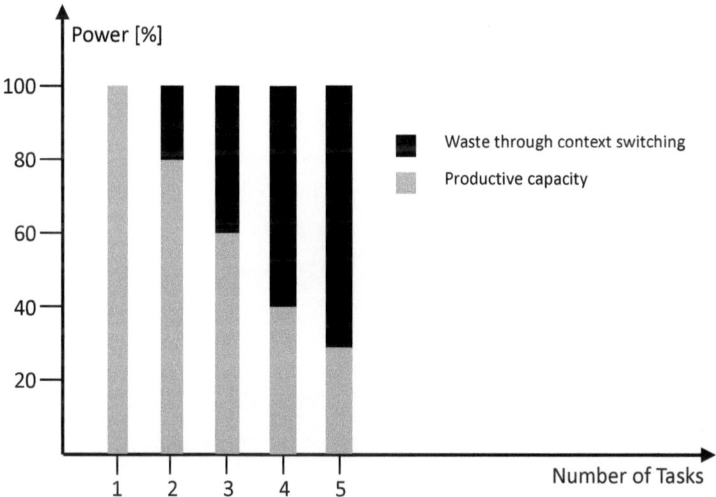

Figure 41: Waste through context switching

The 20 per cent loss per additional task assumed by Weinberg is often the subject of discussion. The specific figure certainly depends on the activity and the frequency of context switches. In product development, however, there are often situations in which complicated considerations have to be made: Designing an algorithm, thinking through a circuit, or constructing a 3D model. In such activities, the brain has to deal with a lot of information at the same time. Many minutes pass before all the required information has been uploaded to the brain and is available in the working memory. You are probably familiar with situations like this in which you find yourself in a state in order to be able to think productively about a particular topic. Any disturbance, even if it is just a short phone call, kills this state and you have to recollect the required information from scratch.

Scrum actively addresses this issue: Focusing on one team and one product makes a significant contribution to team performance by avoiding context changes for everyone involved.

Teams

Team building

Agile teams have no hierarchies and no explicit roles. Nevertheless, implicit roles quickly emerge that make a team work. A common model for these implicit roles comes from the English psychologist Meredith Belbin. He defines nine roles in three categories (Belbin, 1993):

Action-oriented roles
- Shaper
- Implementer
- Completer/Finisher

People-oriented roles
- Coordinator
- Team worker
- Resource Investigator

Celebral roles
- Plant
- Monitor Evaluator
- Specialist

As these implicit roles arise within the team and are not defined by the organization, the path to achieving them is fraught with conflict until communication structures and roles have been established and consolidated. The desired team performance is only achieved once these aspects have stabilized.

The American psychologist Bruce Tuckman describes this life cycle with his well-known model of team building, which he divides into five phases (Tuckman, 1967; Figure 42):

- Forming: The team is newly formed; a spirit of optimism and euphoria prevails. The team's performance exceeds the sum of its individual achievements.
- Storming: Communication structures and role allocations are in flux; the initial euphoria has faded. Conflicts arise, the performance of the team is below the sum of the individual performances. There may also be personnel changes in the team.

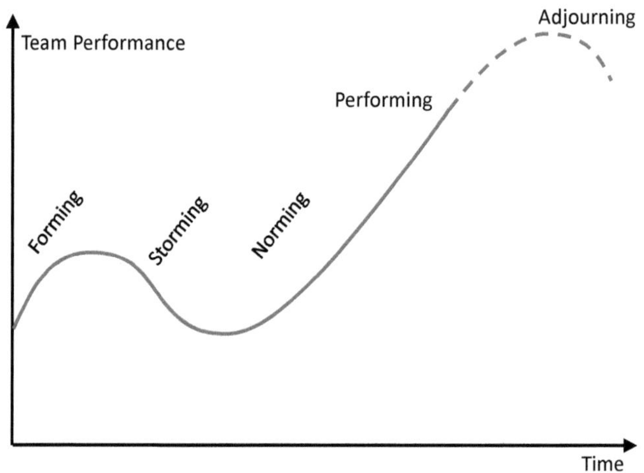

Figure 42: Team performance during team building

- Norming: Communication structures and roles are consolidated, and conflicts are resolved. The team becomes a coordinated organism and team performance increases.
- Performing: An efficient team has been formed, roles and communication structures have been clarified. The performance of the team is significantly higher than the sum of the individual performances.

 People and Teams

- Adjourning (not included in Tuckman's original paper, later added by him): No team is stable in its composition in the long term. Every change, including the addition of new members, starts a new run through the Tuckman phases, starting with Forming.

Scrum relies on stable teams so that – according to Tuckman – they can be maintained in the performing phase. Nevertheless, it is always necessary to make changes to the team composition for social or technical reasons, which temporarily reduces the team's performance as described. So, if a Scrum Team is running towards a tight deadline with its product, the worst approach is to increase the size of the team in order to increase speed. In the medium and long term, team performance may actually increase, but in the short term the team will lose speed due to a new storming phase. In such situations, clearing impediments is the only way to react in the short term.

Co-location

Team-based agile approaches are based on the assumption that all team members are in the same room during their work – such a team is referred to as a co-located team. Customers often question the necessity of this setup, as many organizations are not prepared for such constructs. When it comes to co-location, it is important to know that the performance and ability of a team to perform is largely determined by the communication between team members. If communication is dependent on tools such as telephone or e-mail, the performance of the team inevitably suffers. For example, studies from the 1980s in the aviation industry found a demonstrable link between communication and team performance for cockpit teams (Wiener, Earl, Kanki & Helmreich, 2010).

For me, a co-located team is the only adequate setup for a powerful agile team. If you have to resort to tools, this creates a break in communication and therefore in performance. Whether there is only a door or maybe even an ocean between the team members is irrelevant. A distributed team will never achieve the performance of a co-located team.

Team composition

Team composition, is about the question of which skills or people should be represented in a team. In the Scrum chapter, I described that an agile team must include all the necessary skills to develop, build, and test an Increment. On the other hand, a team should not comprise more than ten people. This means that it is often necessary to use more than one team when developing mechatronic systems.

In settings with several teams, there are now various ways of distributing the required or existing competencies. As a rule, organizations are organized according to competencies. There are departments for design, electronics, software, quality, purchasing, prototyping, personnel, and so forth. These competence silos are shown as vertical bars in Figure 43. The point is that this organizational structure is an attempt to create synergies and increase efficiency. However, some of these competencies are required to develop a product. Products therefore cut across many departments, as shown in the figure by the horizontal bars. If teams are tailored according to the organizational structure, there are many dependencies between these teams – because in order to be able to deliver a functionality in the product, the cooperation of many organizational units is necessary. Such teams are referred to as specialist or domain teams.

Many projects are not traditionally organized according to specialist areas or domains, but analogous to the product architecture: The teams in the project are tailored according to the components of the product and organized in component teams and there are often allocated component managers. In Figure 43, the vertical bars correspond to the components. This illustration also shows that such a structure still requires the cooperation of many teams, even if it is closer to the product than a structure based on domains.

In order to optimize lead times, agile approaches aim to minimize dependencies between teams. From this perspective, it is therefore optimal to set up the teams according to the horizontal bars in Figure 43, i.e., along the product. Such teams are referred to as feature teams, as they are ideally able to develop a functionality – a feature – across all components and specialist areas. They can do this autonomously, without dependencies on other teams. To do this, a feature team must have all the expertise needed to add another functionality to a product increment. This is in line with the expectations of a Scrum Team. Feature teams are therefore cross-functional teams in the same way as Scrum Teams – they combine different skill profiles.

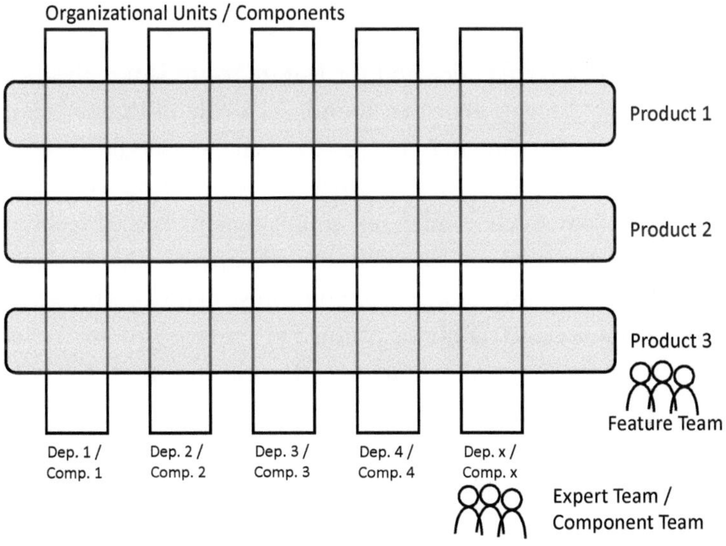

Figure 43: Team composition

The two axes between domain or component on one side and product on the other were previously mapped in matrix structures. The project structure is placed on the organizational structure rotated by 90 degrees. The organizational structure of agile approaches are teams, which is why an agile development organization ideally consists of several feature teams. However, the other axis, that of expertise, must not be lost in the process: Even in agile organizations, subject matter experts must be able to exchange ideas with one another. In contrast to the matrix, however, the feature teams should be the primary structure here and the technical exchange is placed across this structure. In contrast to the matrix, the primary and secondary organizational levels are interchanged. The exchange of expertise is organized through communities of practice or guilds: The experts in a domain work in distributed feature teams and meet regularly in their guilds to exchange expertise and to optimize and harmonize technologies and procedures within their domain across the teams.

In practice, a blanket demand for feature teams is not always expedient; mixed forms are often found. As a rule of thumb, central services and infrastructures, such as test benches, should remain as specialist teams – as should teams that deal with innovative topics or high technology, such as analyses, simulations or special technologies. As far as possible and sensible, the other people should then be organized into feature teams. A good team structure is only created through practice and cannot be planned in a meeting room. As soon as you have the first concept that could work, sprint off. Every few weeks, in the Sprint Retrospective, you have the opportunity to rethink and improve your approaches.

Skill profiles
Many organizations find it difficult to switch to cross-functional teams because a distinct expert culture has developed over time. In

the past, passing on knowledge was often seen as a hindrance and instead of distributing knowledge throughout the entire organization, individual experts were always accessed from different places: A super-expert completed the same work faster than a semi-expert. However, this is only an apparent gain in performance, because in practice, queues form in front of the experts, and these cause pronounced idle times in the course of the project. In addition, dedicated experts generate a significantly higher organizational, coordination and communication effort. On the other hand, agile approaches should not lead to a loss of expertise because – for the sake of flexibility – every team member should be able to work on every task. The ideal is to expand the existing skills profiles and knowledge of experts.

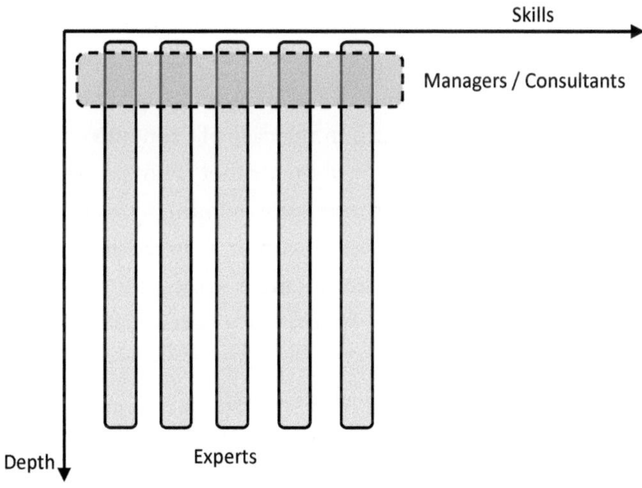

Figure 44: Skill profiles

Figure 44 shows the skill sets of experts as vertical bars.Each profile is very deep, but also very narrow; the profiles do not overlap.

With such profiles, it is difficult for an agile team to work flexibly on the tasks provided in the backlog, as the experts can hardly support one another. An alternative profile is shown above as a horizontal bar: It is the profiles of managers, consultants, and ducks (ducks can walk, fly, swim, but none of them very well). Even with such profiles, it is difficult for a team to develop products because the necessary skills are not developed deeply enough.

The aim in agile teams is to combine the two profiles described above: The expert status in one specialist area is maintained and at the same time basic knowledge in the other domains is acquired. This results in a T-shaped skill set (Figure 45). How pronounced the crossbar of the T is depends on the product and industry. A software developer will certainly never model the 3D model for a housing and a mechanical designer will never review the EMC concept of a printed circuit board. However, the minimum requirement for the crossbar is that the team members understand the topics and problems of the others to some extent, and can also provide information on some aspects from within the team. In the case of pronounced crossbars, the team members can support each other with simple tasks. More is often possible here than those involved initially assume. For example, the evaluation of a test bench run in a spreadsheet can be taken over by almost any other team member after a short briefing, therefore relieving the test bench expert if they become a bottleneck in the current iteration.

So how can you achieve such T-shaped skill sets in your teams? These profiles can neither be trained nor designed, they will emerge through teamwork. Pair doing, i.e., working together on a task, makes a valuable contribution to this. This originally comes from software development and the extreme programming movement. It has proven its worth when two people work on difficult programming tasks in front of one computer. This has the following advantages: The field of vision when working on the task is expanded and shared. The person operating the PC has the tactical view and ensures that the commands are entered correctly. The person sitting next to them has the strategic view: Which other modules are affected by the current action? What interactions are there with the system architecture? At the same time, a review of the work takes place and knowledge is exchanged.

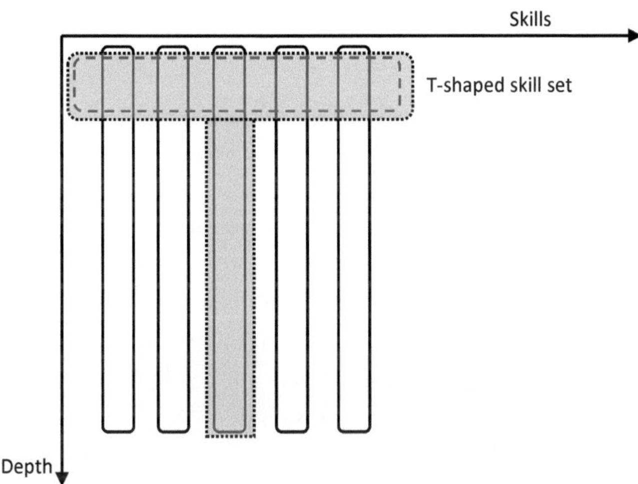

Figure 45: T-shaped skill set

You can also use this approach for any other activity outside of software development. Practice has shown that working together brings so many advantages in terms of solution and work quality that the assumed waste of pairing is more than compensated for. The parallel execution of different activities demanded by traditional efficiency thinking is much more likely to generate technical debt in terms of implementation quality, innovation, and knowledge sharing when it comes to difficult tasks. Encourage your team members to work with several people on one task. This is the way to flexible and fast cross-functional teams.

Bringing agility to the organization

This book cannot provide a comprehensive overview of all elements of agile transformation and organizational development. Therefore, I offer a brief overview of the interrelationships and challenges as well as an example of an agile approach to organizational development.

Agile transformations

Agile transformation generally refers to the spread of agile approaches in an organization, i.e., the *agilization* of the organization. The major challenge compared to converting a single team to Scrum is the significantly greater complexity and the more diverse mix of people involved.

For an individual team, a single manager can create a safe space for learning and form the interface with the previous organization. Ideally, enthusiastic employees who participate voluntarily are also deployed for such pilot teams. If these approaches are now to be rolled out more broadly, a complex social and economic structure is involved in which not everyone is an enthusiastic agilist.

Planning an agile rollout and mandating work with Scrum & Co. rarely works in practice. People cannot be forced to change; they can only change on their own initiative – otherwise they will resist or leave the company. One possible model for tackling such changes together with employees is OpenSpace Agility, which I will briefly introduce in the next section. A rollout as described above does not only fail due to a lack of voluntariness. Organizations are complex environments and a plan-driven approach with fixed deadlines will usually fail to achieve the goals.

For agile transformations, it makes sense to use the agile mindset of complex environments for agile transformation. But how can Transparency, Inspection, and Adaptation be used for organizational development? As in product development, transparency involves dealing honestly with the current status and what is and is not possible. Consequently, we also need transparency about what people want and are able to do. Without wanting it, there can be no change. The aim is therefore to invite people to change. An invitation always includes the opportunity to say no. Who then goes onto accept the invitation and who doesn't? This is the first piece of transparency for agile transformation.

The invitation is only the beginning, the agilization of the organization should be divided into small steps and the plan should be adapted after each step. The Scrum mechanisms apply here on a large scale: People plan a timebox with experiments on agile working methods and then meet to discuss the results and plan the next timebox. All levels must follow this path together, from top management to developers. Everyone will fail at some things at some point, and succeed with others at another point. The result is a learning organization.

Example: OpenSpace Agility

OpenSpace Agility (OSA) is a tried-and-tested model by Daniel Mezick for invitation-based iterative-incremental organizational development (Mezick et al., 2019). Recurring open space events form the core of the model. Open Space is a format for invitation-based, self-organized large group work in the context of organizational change. OSA sets any number of iterations – learning chapters – in succession, not limited by an open space event in each case. Open Spaces instigate ideas and collaborators hidden within the organization that would remain hidden if change were forced.

The invitations to the Open Spaces and to change are extended by top management. Top management and all managers in the company must back the change and support it. If this cannot be achieved or if there are concerns regarding transparency, voluntariness, and self-organization, the process is aborted. The process can also be terminated after the first open space event if too few people have accepted the invitation and the experiments cannot be started. In both cases, it becomes clear that the organization is unable or unwilling to tackle this change. In order to move forward, managers must now create awareness of why the change is necessary. So, it may well be that after a year of storytelling, another invitation to an open space event is extended and a new attempt is made.

It is important to note that going into a change with open space events does not mean anarchy. It is a democratized suggestion system; the management sets the goals for the change and still decides what is done and what is not.

Figure 46 shows the basic process of OpenSpace Agility. In a preparatory phase, managers are prepared for the change through

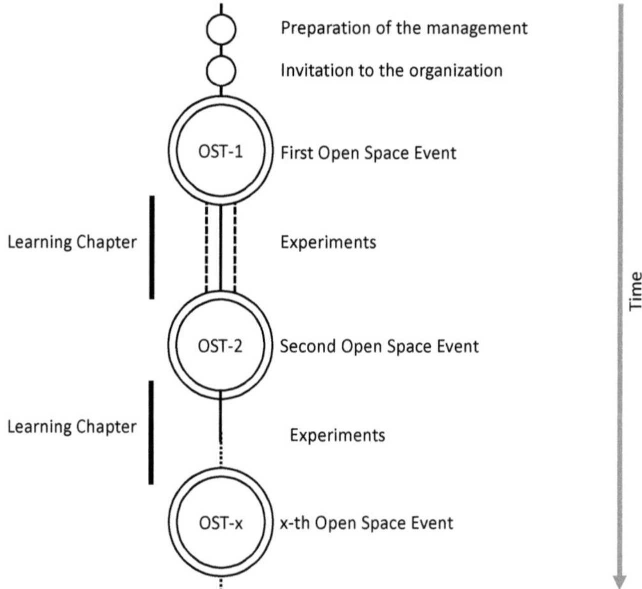

Figure 46: OpenSpace Agility process (schematic)

coaching and small workshops. Employees are trained in agile topics so that goal-oriented, high-quality discussions can be held in the first Open Space event. At the same time, the goal of the Open Space is formulated and the people affected by the change are invited. This may well involve several thousand people.

In the first Open Space, experiments are defined that can be used to determine how agile working methods can be used in the organization and which of them are effective. The Agile Manifesto provides the framework for the experiments. Whether Kanban, Scrum or other approaches are suitable is determined by the experiments and not by a preliminary definition by management.

The first phase of experimentation is accompanied by external coaches to provide support during the change process. One result of the Open Spaces are recommendations to top management, which they must decide on promptly (within a few days) so that the experiments can begin. These are usually decisions on structures and investments.

In order for the experimentation phase to take place, the right people must of course be present in the Open Space. If they do not accept the invitation, the change stops at this point. No one is forced to adopt the new working methods.

In the second Open Space, the findings from the experiments are evaluated together and the approaches – which can be used immediately in everyday life – are defined. The second learning chapter begins, in which the number of experiments is reduced, as the first aspects have already been incorporated into daily work. However, this time is also another learning chapter and ends with the next Open Space. In this way, a learning chapter follows a learning chapter, and a learning organization is created. After the second Open Space, the external coaches end their assignment. This is announced to everyone involved right at the start of OSA so that no one is tempted to transfer responsibility to external coaches. It also gives the organization a sense of achievement after the first experimental phase if it can take the first small steps towards agility on its own. If necessary, other coaches can be brought in for the second learning chapter. Thanks to the high degree of self-organization and personal responsibility, organizations can cope with OpenSpace Agility at a fraction of the cost of external consultants.

Literature

Arnold J. & Yüce, Ö. (2013). Black Swan Farming Using Cost of Delay: Discover, Nurture and Speed Up Delivery of Value. In *Proceedings of the 2013 Agile Conference*. Washington: IEEE Computer Society.

Beck K. et al. (2019): *Manifesto for Agile Software Development.* Verfügbar unter http://agilemanifesto.org [06.01.2019].

Belbin, M.R. (1993). *Team Roles At Work.* Oxford: Butterworth Heinemann.

Cohn, M. (2005). *Agile Estimating and Planning.* Amsterdam: Pearson Education.

Coplien, J. (1994). Borland Software Craftsmanship: A New Look at Process, Quality and Productivity. In *Proceedings of the 5th Annual Borland International Conference*, June 5, 1994. Orlando.

Furuhjelm, J., Segertoft, J., Justice, J. & Sutherland J.J. (2017). *Owning the Sky with Agile: Building a Jet Fighter Faster, Cheaper, Better with Scrum.* Verfügbar unter www.scruminc.com/wp-content/uploads/2015/09/Release-version_Owning-the-Sky-with-Agile.pdf [06.01.2019].

Goldratt, E. & Cox, J. (1984). *The Goal.* Great Barrington: North River Press.

Larman, C. & Vodde, B. (2008). *Scaling Lean & Agile Development Thinking and Organizational Tools for Large-Scale Scrum: Successful Large, Multisite and Offshore Products with Large-scale Scrum.* Amsterdam: Addison-Wesley.

Larman, C. & Vodde, B. (2010). *Practices for Scaling Lean and Agile Development: Large, Multisite, and Offshore Product Development with Large-Scale Scrum.* Amsterdam: Addison-Wesley.

McGonigal, J. (2012). *Reality is Broken: Why Games Make Us Better and How They Can Change the World.* London: Vintage.

Mezick, D. (2012). *The Culture Game: Tools for the Agile Manager.* FreeStanding Press.

Mezick, D. et al. (2015). *The OpenSpace Agility Handbook.* New Technology Solutions.

Morgan, J. & Liker, J. (2018). *Designing the Future.* New York City: McGraw-Hill Education.

Ohno, T. (1988). *Toyota Production System.* New York: Productivity Press.

Pfeffer, J. & Sasse, M. (2020): *OpenSpace Agility compact.* Wangen: peppair.

Pink, D. (2009). *Drive: The Surprising Truth About What Really Motivates Us*. New York: Riverhead Press.

Reinertsen, D. (2009). *The Principles of Product Development Flow: Second Generation Lean Product Development*. Redondo Beach: Celeritas Pub.

Schwaber, K. & Sutherland, J. (2020). *The Scrum Guide: The Definitive Guide to Scrum*. Verfügbar unter www.scrumguides.org [06.01.2022].

Schwaber, K. (1995). *SCRUM Development Process*, OOPSLA Conference 1995.

Schwaber, K. & Sutherland, J (2011). *The Scrum Papers. Nut, Bolts, and Origins of an Agile Framework*.

Snowden, D.J. & Boone, M.E. (2007). A Leader's Framework for Decision Making. *Harvard Business Review*, November 2007.

Stacey, R. (2000). *Strategic Management and Organisational Dynamics: The Challenge of Complexity* (Third edition). London: Pearson Education.

Sutherland, J. (2014). *SCRUM: The Art of Doing Twice the Work in Half the Time*. New York: Crown Publishing.

Techt, U. (2010). *Goldratt und die Theory of Constraints: Der Quantensprung im Management*. Moers: Edition La Colombe.

Takeuchi, H. & Nonaka, I. (1986). The New New Product Development Game. *Harvard Business Review*, Januar 1986.

Tuckman, B. (1965). Developmental sequence in small groups. *Psychological Bulletin*. 63, 1965, 384–399.

Ward, A. & Sobek, D. (2014). *Lean Product and Process Development*. Boston: Lean Enterprise Institute.

Weinberg, G.M. (1992). *Quality Software Management Volume 1: Systems Thinking*. New York: Dorset House Publishing.

Wiener, E., Kanki, B. & Helmreich R. (2010). *Crew Resource Management*. Boston: Academic Press.